GLOBAL HIVE

GLOBAL HIVE

What the bee crisis teaches us about building a sustainable world

Horst Kornberger

Floris
Books

First published as *Global Hive* by
the School of Integral Art in 2012
This edition published by Floris Books in 2019

 Also available as an eBook

British Library CIP data available
ISBN 978-178250-569-3
Printed and bound in Great Britain
by Bell & Bain, Ltd

CONTENTS

ACKNOWLEDGEMENTS

Much of the content of this book was inspired by *The Bee Master*, a four-day festival drama written by Jennifer Kornberger in collaboration with West Australian composer Paul Lawrence. The play contained in metaphoric form many of the themes elaborated in these pages. Directing *The Bee Master* became the immediate starting point for writing this volume.

Friend and artist collaborator Tom Müller invited me to contribute to the TIME – EMIT exhibit in Fremantle in 2001, which sparked my first bee-related artwork. He also helped with the first, self-published edition of this book.

I am deeply indebted to friend, colleague and supporter Ann Reeves for ongoing encouragement, important conversations and invaluable editing advice. I am extremely grateful to Janet Blagg, friend, neighbour and master editor who has again given generously of her time, expertise and insights to bring this work into the world. Christopher Moore gave valuable advice on the structure of the book.

I am indebted to biologist and bee master Johannes Wirz for crucial conversations and very helpful comments; to Michael Spence for articulating complex economic concepts in down-to-earth language, and to Brian Keats, Guenther Hauk, Craig Holdrege, Matthew Barton, David Heaf, David Adams and Michael Weiler, all of whom contributed important insights. Particular thanks to John Stubley and Katie Dobb for many helpful conversations on the topic of social change.

My thanks to Alicia Braxton, who has generously supported the writing of this book.

An earlier version of the material on the bee queen appeared in the anthology *Queen of the Sun*, compiled by Taggart Siegel and Jon Betz.

1

FROM BEUYS TO BEES

When written in Chinese the word 'crisis' is composed of two characters – one represents danger, and the other represents opportunity.

John F. Kennedy

We are becoming used to environmental disasters. The bee crisis, however, strikes a singular chord. Everyone is alarmed because everyone is affected. One third of the world's food production depends on bees. The potential loss of pollinating hives is a clarion call even to environmental sceptics, the loss of biodiversity a frightful prospect to those with ecological concerns. A growing world population and a declining number of bees is a dangerous mismatch.

The first signs of colony collapse surfaced in Texas and Louisiana in 1960. From 1972 feral bee populations began to fall in the United States and are now almost extinct. Trachea and Varroa mites started to wreak havoc in colonies in the 1980s. France reported major unexplained losses of bee populations in 1992. By 2005 fifty per cent of domesticated bees had perished in the US. The United Kingdom became affected in 2006. In the same year the term Colony Collapse Disorder was coined to describe what was rapidly becoming an agricultural catastrophe. Two years later many US beekeepers reported losses of up to ninety per cent. Belgium, the Netherlands, France, Greece, Italy, Portugal and Spain suffered substantial declines, and other European countries soon followed. In the meantime, severe colony collapse has struck China and Japan. Egypt is also affected.

Beekeepers are at a loss as to how to stem the tide of destruction. Scientists have begun a frantic search for causes. Many accuse the Varroa mite, but this mite coexisted with Asian bees long before colonies began to collapse. Other researchers blame monoculture, genetically modified crops, herbicides, pesticides and electro-smog. No doubt these all play their ignominious part. But bees are dying in areas unaffected by herbicides. Colonies collapse where there is little electro-smog.

In this book I explore the reasons behind the bee crisis and suggest a new way of thinking about it. My aim is not to discredit current trends of thought but to establish communication between different ways of seeing the world. I believe that the gap that separates these ways of seeing is the same into which species after species disappear.

I am not a professional beekeeper, but I have been marginally involved in amateur apiculture in Western Australia. My own relation to bees was initially an artistic one. Inspired by the artist Joseph Beuys, I worked with honey as an artistic medium in search of new icons for the environmental age. Beuys used honey, wax, fat, felt and copper as metaphors for social transformation. My own work led me to creations like *Honeyclock*[1] and *Buddha in Honey* (see Chapters 15 and 24), a series of visual meditations on compassionate ecology. These, together with *The Bee Master*[2], a contemporary festival drama written by Jennifer Kornberger, were part of the initial momentum for this book.

During this work I came across a lecture series on bees from 1923 that influenced much of Beuys' practice, particularly his social vision. The same pages that sparked Beuys' social sculpture opened a door to me to a new compassionate ecology. What I read fuelled my passion for bees and fused with my interest in understanding paradigms: the way they come into being, establish themselves and claim more and more minds until they rule with singular power over decades, centuries and sometimes millennia. The transcript of the lecture series shed light on bees and paradigms simultaneously. It spoke of interconnectedness, wholeness and of ways to learn from the future before it becomes a mistaken past.

The lectures that inspired Beuys predated the bee colony collapse by some eighty years. The speaker was neither a beekeeper nor an expert in the way we understand the word, yet the future demise of the honeybee was predicted in his first lecture and its causes clearly outlined:

> But now we come to this whole new chapter concerning the artificial breeding of bees... Much can be said for the artificial breeding of queen bees, because it does simplify things. But the strong bonding of bee generations, a bee family, will be detrimentally affected over a longer period... In certain respects you will be able only to praise artificial breeding, if all necessary precautions are taken... But we will have to wait and see how things will look after fifty to eighty years. Certain forces that have operated organically in the beehive until now will become mechanised... It won't be possible to establish the intimate relationship between the queen bee you have purchased and the worker bees the way it would arise all by itself in nature. But at the beginning the effects of this are not apparent...
>
> This is the way things stand... with the artificial breeding of bees. You can increase their production of honey, all the work they do, and even the worker bees' capability of accomplishing this work... We'll see that what proves to be an extraordinarily favourable measure upon which something is based today may appear to be good, but that a century from now all breeding of bees would cease if only artificially produced bees were used. We want to be able to see how that which is so wonderfully favourable can change in such a way that it can, in time, gradually destroy whatever was positive in this procedure.[3]

In the audience sat a conventional beekeeper called Mr Mueller. He represented the state of the art of beekeeping: the professional expert. Of course, this practical, knowledgeable man was very much taken aback by the speaker's assertions, an understandable reaction given that the speaker was indicting the very practices that made beekeeping profitable for a catastrophe nobody in the room would live long enough to see. Nor could Mueller see any evidence for such a catastrophe. Later in the series the speaker, referring to heated contributions from Mr Mueller, reiterates his position:

> Nevertheless, it is important to gain this insight – that it is one matter if you let nature take its course and only help to steer it in the right direction when necessary, but it is entirely another matter to apply artificial methods to speed things along. I really don't want to take a strong position against what Mr Mueller has stated. It is quite correct that we can't determine these matters today; it will have to be delayed until a later time. Let's talk to each other again in one hundred years, Mr Mueller; then we'll see what kind of opinion you will have ... [4]

Reading this passage, I can almost hear the laughter in the room – but today the audience would not join in. Hives are dying worldwide and the plight of the honeybee has become dire reality. If it continues, honey is the least we will lose. Bees are the agronomists of nature, ecologists in action. They pollinate our major crops and maintain biodiversity. An estimated one third of the food staples on this planet is dependent on their ceaseless toil. Bees are major collaborators we cannot afford to lose.

The lecturer on bees was one of the few great universalists of the last century, Rudolf Steiner, perhaps best known as the founder of Waldorf education and biodynamic farming. While I am fascinated that Steiner was able to foresee the bee collapse, explain its causes and accurately estimate the timescale, I am even more fascinated by the fact that his contribution is ignored by current researchers. I see this as the one-sided orientation of a mindset that has had its time. There is something in the present paradigm that would rather suffer a crisis, no matter how devastating, than entertain the validity of a different way of seeing the world. This is as important a problem as the bee crisis, deforestation, pollution and global warming together, for it points to the cause behind all these: the current scientific paradigm, with its default position that solutions are found in details and that parts make the whole. This mindset applies the microscope of the mind where the 'macroscope' of applied imagination should prevail.

Steiner's lectures are confronting in this respect. They oppose prevailing trends with a radical reimagining of our concept of ecology. Much of Steiner's lecture cycle will appear arcane to those unfamiliar with his work; however, his basic insights about bees and why they will disappear are perfectly intelligible to common sense as well as to compassionate sensitivities we all already have, but rarely apply. While it requires an expert to manage an apiary and trained scientists to fathom the chemical composition of queen substance, the most important insights about bees are well within our reach – as are the reasons they disappear.

My first aim in this book is to facilitate this understanding of *Apis mellifera*, my second to mediate between differing paradigms. Without cross-pollination of mindsets, ecology remains infertile. A new beehive of paradigms is needed to solve the plight of bees and the current problems of nature.

2

HONEY HUNT

If there is a buzzing noise, somebody is making a buzzing noise, and the only reason for making a buzzing noise I know of is because you are a bee... and the only reason for being a bee I know of is making honey.

A.A. Milne, *Winnie the Pooh*

Like Pooh Bear, we think of honey when we think of bees. This is as natural to us as it is detrimental to bees. The worldwide colony collapse reminds us that bees have an ecological function more crucial than simply satisfying our palate. Yet it was honey that led to beekeeping. Bees produce what we crave, yet they accumulate honey to secure their future. They feed their larvae as well as themselves on the floral extract. Their stores need to be greater where winters are long and summers short, and they have to have enough to survive when spring comes late, when summers fail, when calamity befalls the hive or when their population increases.

Honey is far more intimately related to bees than nuts are to squirrels or seeds to chipmunks. The small floral ambassador, visiting thousands of flowers every day, ingests the nectar and so unites herself with the flowery syrup. Honey is always part bee. Back in the hive she regurgitates her precious load and deposits it in hexagonal chambers of wax. There she kneads it with her long proboscis until the water content evaporates. When perfect consistency is reached, bee chemists add a minute proportion of their poison for durability, a homoeopathic dose of bee sting, like a spice, before sealing the chamber.

Honey is a whole landscape condensed into sweetness. It is flowers made fluid and seasons distilled. A slab of comb may contain the lavender fields of Provence, fragrant alpine meadows in August, the bloom of the Sidr tree in sunblazed Yemen. Honey is environment made into taste. It is the final perfected result of the widespread, ceaseless, beneficial activity of bees in the service of nature: honey is ecology in liquid form.

To ancient humanity honey was a sacred substance. Far more than bodily nourishment, it was a divine gift condensing the diluted sweetness of the world into thick liquid gold. It was symbolic of everything precious, refined, rare.

From the dawn of humanity men and women have hunted the sweet syrup, braving danger and pain to wrest the prized substance from its

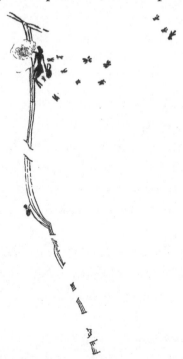

Figure 1. *Bicorp Man*. Rock art, *c.* 6000 BCE, Araña Caves, Valencia, Spain.

watchful guards. Equipped with nothing but smoke and daring they have thrust their hands into small holes and cut comb from the insides of humming caves (see Figure 1). They have risked and sometimes lost their lives in the pursuit of sweetness. When hunters and gatherers settled, hives were housed too. Farmers and bees have much in common: both are settlers and agriculturalists that transform nature.

Both work hard, harvest and store their surplus for times of need. Both form complex societies structured by division of labour. The proximity of hive to farm is both convenient and symbolic. It is the beginning of a symbiosis that has never ceased: agriculture and apiculture belong together. The sweet syrup, though highly prized, is only one of bees' manifold blessings. Consider their collaboration in all matters agricultural.

Bees pollinate flowers, herbs, vegetables and fruit trees and so supplement the farmer's efforts with their own. Gardens grow more abundant in proximity to a hive. A fruit tree visited by bees bears more fruit. Some varieties of fruit grow larger when visited by more than one bee. Nature thrives wherever there are bees.

Since the domestication of bees, the histories of *Apis mellifera* and *Homo sapiens* have been intimately related. Beekeeping has changed with time, most radically over the last hundred years. We have added science to tradition and efficiency to time-honoured rhythms, sometimes to the benefit, sometimes to the detriment, of hives. Bees are exceptional creatures. They remain part of nature even in captivity and are never as fully domesticated as dogs or cows. Because of this, beekeeping holds up a mirror to our understanding of nature, and is an exacting ecological thermostat for the standard of our environmental care. The first beekeepers kept their hives in hollow logs or cylinders of clay (see Figure 2, overleaf), imitating traditional bee habitation in hollow trees and caves. Although many of these hives were destroyed when honey was harvested, this did not impact the overall health of the bee population.

Intervention increased as cultures evolved. The ancient Egyptians

Figure 2. *Beehives in the Tomb of Pabasa*. Tomb painting, *c*. 2400 BCE. The beekeeper next to the stacked clay hives may be the earliest depiction of beekeeping.

moved their bees according to season, fixing their hives on river rafts and floating their colonies southwards along the calm waters of the Nile as spring unfolded. The Greeks began to use the convenience of top bars (removable bars on which the comb is hung, which allow easier access to honey). The Romans were the first to cast a purely commercial eye on the once sacred bees.

In early medieval times apiculture survived in monasteries. Monks treasured their apiaries for their gift of honey, the only sweetener at the time, and as sources of beeswax, indispensable to the rituals of the church. Around 1400, fragile pottery hives were replaced by straw baskets in Europe, and in 1650 the practice of beekeeping changed radically with the introduction of octagonal wooden hives and the 'super' – a structure that expanded the hive and made space for new comb that contained no brood. When filled with honey these supers could be removed without destroying the colony. Now honey could be continuously harvested and hives remained alive. Apiaries expanded and yields increased.

The age of industrialisation could not but affect the industrious bees. When English entrepreneurs pioneered the profitable running of factories, Lorenzo Langstroth introduced similar rationales to beekeeping in the US. He applied pioneering apiarist François Huber's discovery of 'bee space' – the distance that bees keep between their combs – to the design of efficient rectangular hives fitted with a succession of frames. These frames made bees adjust their rounded combs to the convenience of their keepers. Combs would no longer bond to the box and could be easily removed. The new hive design allowed for safe control of the brood and convenient access to honey.

In 1857 Johannes Mehring further perfected Langstroth's design by inserting artificial, preprinted hexagonal wax foundations into rectangular frames and so regimented the natural architecture of comb. Nine years later Franz Hruschka invented the honey extractor, introducing machinery for the first time. Beekeeping was becoming commercially viable.

Yields peaked at the beginning of the twentieth century with the artificial breeding of queens. A few decades later, artificial fertilisation of queens surpassed all previous interventions. There was now more control over the process, more convenience, more honey for more people and, of course, more profit. But an inviolable line had been crossed and *Apis mellifera* was being compromised to the point of extinction. This was the point Steiner was making in his lectures in 1923. What began as collaboration became manipulation and coercion and, finally, subjugation. We need to examine this line if we want to safeguard our bees. On one side of this line lies sustainable collaboration. On the other is the ever-growing gap into which species disappear.

3

DOMESTICATION AND WHAT IT MEANS

The greatness of a nation and its moral progress can be judged by the way its animals are treated.

Attributed to Mahatma Gandhi

Domestication means different things to different animals. When wolf cubs were reared into dogs, the shepherds' fiends became their trusted friends. Though taken from nature into culture they did not lose their essential features. Wolves are no loners; they live and hunt in packs. The initial bond to their own kind is replaced by one to their master. We say those wolves are tamed, but their courageous fierceness is not extinguished, as every valiant watchdog testifies. The dog's predatory instincts have served hunters through the ages, and their keen sense of smell has many applications. And though not all dogs apply all these gifts, the dog in all dogs does; enough of their essential nature is engaged to make life agreeable.

Cats too are well adapted. With the exception of lions, who enjoy the royal prerogative to differ, wildcats are loners. They meet to mate but otherwise enjoy the solitude of their haunts. When tamed, they retain this trait. Even in the jungles of suburbia they stay aloof. They hide away, enigmatic, even mysterious. They like to come and go, and grace their infatuated owners only occasionally with their dignified company. Domestic cats retain an aristocratic attitude that differs markedly from that of the ever servile, tail-wagging dog. In short, the cat has retained much of its independent cat-ness and is, as anyone can see, greatly at ease with itself.

The story of the chicken is very different. To take free-roaming fowl from the undergrowth of tropical forest and cramp them into a working camp for the mass production of eggs is a violation of their nature. To confine their once free movement, deprive them of their curious, scratching, clucking search for food and fill them with artificial staple goes against the grain of their needs.

Dogs and cats are lucky in that they do not produce anything that can be exploited. Chickens do. So do cows. Cows are herd animals. Domesticated and kept between generous fences, this animal will gladly do what it has always done: amble through the meadow and chew its food. Cows are a kind of miraculous laboratory that transforms poor staple into rich, warm, creamy milk. When I think of milk I see a vast river of nourishment streaming from the cow in all cows into the world, dividing itself into cream and butter and milk and yoghurt and cheeses as diverse as the landscape from which these products come. But cows give more than milk. This motherly animal offers every part of herself as gift: since time immemorial her hide has been turned into leather and her intestines into string. Her bones are made into gravy and her meat into food. Horn and hooves, once used as ornaments, gain new importance in biodynamic farming. Her dung is especially valuable: nothing rejuvenates soil as much. Meadow appropriately grazed by cows will become more fertile. There is no other animal as universally beneficent as the cow.

To cramp cattle into tight feed lots, force them to eat grain instead of grass and breed them to yield more milk than they are willing to give stretches their generous nature to the limit. It is painful to imagine this slow, heavy, utterly phlegmatic animal continuously administered hormones to maximise milk yield. A hormone is to the cow body what a deadline is to a writer: an incentive to produce more in less time. The result is an enormous degree of stress. And while such stress may occasionally be appropriate in human life, it is not for a cow, whose very essence is the opposite of frantic activity. To impose substantial stress on a cow means nothing less than to drive her essential, slow, rhythmic nature to extinction.

Here, as with the bees, we overstep the mark, and encroach on essential cow-ness. This abuse *ad absurdum* eventually rebounds. The result is the worldwide increase in cow diseases and a fall in the quality of milk. Forced overproduction of milk in one part of the world topples the balance in another: dairy surplus cheaply dumped on developing countries ruins vulnerable agrarian communities. Considering the scale of violation, the damage to the cow is remarkably minimal, perhaps because the cow in all cows is not in one place: there are still millions of cattle roaming in pampas and prairies or grazing on bell-tolled meadows in the Alps. Maybe the two hundred million cows freely ambling the streets of India help keep the balance. I like to think of all those four-hoofed temples of patience, stopping the traffic and slowing the frantic hassle of life to a digestible speed. Animals are adaptable. Treated according to their nature, they will thrive under very different conditions. If, however, their needs are ignored, compromised or negated they will respond with weakness, illness or, in the case of bees, a worldwide collapse. It is crucial therefore to know the essential nature of an animal and work with it rather than against it. What, then, is the essential nature of bees?

4

THE BEE IN ALL BEES

Boy: Vyasa, why did my family murder one another?
Vyasa: Because they forgot what is essential.

The *Mahābhārata*

Bees are insects closely related to wasps and ants. There are more than 20,000 species of wild bees, many of them solitary, while others gather in small and sometimes only temporary colonies. The domesticated bee, *Apis mellifera*, forms large and lasting colonies containing up to 60,000 sterile female workers, a few hundred male drones and one queen mother dedicated to the rearing of brood (Figure 3, overleaf). All bees in a colony are her offspring. The queen lives up to five years, while workers die after several weeks or months, depending on the season. All bees have two transparent wings, six legs, and a body clearly divided into three segments: head, thorax and abdomen. Domesticated bees elaborate on this basic model with a number of ingenious additions. Tufts of hair continuously gather pollen. As the bee flies from one flower to the next she sweeps her forelegs over the pollen dust and combs it to her hindquarters, which are equipped with pollen-pressing and -gathering 'baskets'. She has also developed a long proboscis-like tongue, spoon-shaped at the end to scoop nectar from tubular flowers, and a special stomach in which to store it. Female workers are armed with a sting and are able to exude beeswax from their abdomen.

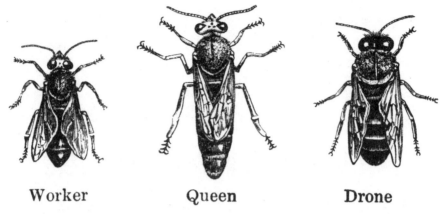

| Worker | Queen | Drone |

Figure 3. Types of bee.

The sophisticated organisation of the worker bee is matched by an equally complex social organisation of the hive. And it is by contemplating the latter that we can approach the bee in all bees.

The most obvious thing about bees is their sociability. Relatedness is part of their nature, and community the essential theme of the hive. Birds live in flocks, wolves in packs, cattle in herds. But a single bird will survive in a cage, an isolated wolf can fend for itself, a cow on its own will not die of loneliness – but a bee will. It cannot be separated from its kind, and least of all from its queen. Even if one tends to an individual bee's every need and gives it ample space, water and feed, it will still perish. For the bee, sociability is more important than water. It is not only honey that keeps bees alive, but their dedication to hive and queen.

The single bee cannot live without the hive. The hive cannot exist without the queen. Take the queen away and prevent the rearing of her successor, and the whole hive will perish. The intricate city state, the rich stores of honey, the complex rituals of the workers: the whole finely tuned and highly organised civilisation will fall apart, despite abundant stores of honey and thousands of workers. Without the keystone of the queen, the architecture of the hive falls apart. But if the queen is restored at any point, even when most of the hive has perished, the colony will latch onto life.

We need not be experts to appreciate the bond between bees and their queen. Life is worthwhile for the bees as long as their queen exists;

it ceases to have meaning without her. The essential is what bees cannot live without: the bee in all bees is its relatedness, the all-pervading familial connection that links bees to their queen.

The queen's life is one of continuous labour. Her sole task is the generation of brood. A queen may lay up to 1,500 eggs a day. Every minute, a new bee issues from her creative capabilities. She rules her offspring through 'queen substance', a scent she emanates from her head glands. By means of queen substance the queen spreads herself over the entire colony, regulating the social life of her subjects. Queen substance is both a bonding agent and a royal decree, communal assurance and communication device. It is the material manifestation of the relatedness of bees. The queen's ladies-in-waiting lick this precious scent from her body and spread it throughout the queendom. When the workers sense this scent, they know that the queen is in place, the hive is secure and the future cared for.

Every queen has her own scent. If a queen raised in another colony is artificially introduced into the hive, bees sense her alien scent and are hostile. The new scent reeks of strangeness, usurpation, of a foreign and superimposed rule, an unknown style of leadership. It takes time and a little 'persuasion' by the beekeeper to make them accept the new matriarch. Persuasion here means to confront the hive with its own extinction: killing the old queen and removing all brood. With no more eggs, no more bees; with no more bees, no more hive. The bees are left facing the death of their colony. It is only after they have been under the threat of impending destruction for a while that the beekeeper will place a new queen into the hive.

She is the bees' only hope for survival in a time of utter distress. They will eventually accept her, but not immediately. For this reason, the queen is placed in a cage when she is put in the hive for the first time. Even faced with death, it takes time for the bees to accept the new ruler. But accept they will, for without a new mother replenishing their diminishing ranks they are doomed. The new queen will breed bees for the hive, and the hive will make honey for its keeper. The work will go on. The hive will survive. All will appear to be bee business as usual. For a time.

5

THE CAGED QUEEN

For bees enough honey is enough. For the beekeeper there is never enough.

R. Chapman Taylor and I. Davey, *Practical beekeeping*

Until the beginning of the twentieth century, queens bred within their hive; homegrown royalty ruled over a population of royal descent. Native queens lived to supply their hives for many years, and if a queen failed, another replaced her from the same stock. Until a hundred years ago the social fabric of the hive was never broken. Queens mated in a time-honoured ritual with a number of regional drones, who thus added male variety to the pure female line (more on this in Chapter 9).

In modern beekeeping, a queen is routinely killed and replaced after a year with a specially bred queen. The rationale behind breeding queens is to select bee mothers with advantageous traits and mate them with a select population of drones for the purpose of producing extremely diligent bees giving high yields of honey. For the reason of profit alone, beekeepers interfere with natural mating habits that once ensured a wide variety of traits among individual bees coming from a wide variety of male drones. This genetic diversity regulated by local hives has become compromised by short-term economic imperatives.

Increasing yields are achieved at the cost of breaking the bond between queen and bees. The bees' lot is to be continually orphaned, forcefully conscripted under a new and foreign rule to serve a usurping matriarch governed by a beekeeper.

The formula of killing old matriarchs in favour of new queens seems to work, hence queen breeding has become a thriving business of its own. Reared to maturity, the young matriarchs-to-be are put into cages and sent to apiaries all over the world, bringing not only the scent of a new queen, but also that of a faraway hive mixed with the strange emanations of paper and tape and transport by air and land: all anathema to scent-sensitive bees.

Nature is patient, but not forever. Bees can adjust, but not to everything. We ask too much of them. Beekeepers feed them on sugary syrup instead of honey. They patiently work artificial wax foundations instead of their own comb. They fit their curved architectural artworks to prefabricated rectangular frames. All this compromises the native self-expression of bees. But the imposition of foreign queens is interference on another level. It affects the most essential feature of the bee: the ancient bond between queen and her hive. Every time a foreign queen is forced onto a hive, continuity ceases. She has no part in the history and biological tradition of the hive and so the essential nature of the hive is lost. The result is the slow-acting poison affecting the connective tissue of hives worldwide – the global immune disease of bees that we call colony collapse. Hives die when the bee in all bees is violated, as it has been for a century.

Immunity is a function of good health. When the immune system is functioning well an organism is able to thrive according to its own laws. Everyone is exposed to cold viruses, but not everyone gets the flu. And few of those who get flu will die from it. If, however, immunity is severely compromised, then even a harmless cold can cause death.

To sum up once more: bees are creatures of connectivity. The single bee is intensely bonded to the hive, and the hive to the queen. This relatedness is an expression of the bee in all bees, of what matters most to the hive. To continuously break this relatedness is an attack against the essential bee, the overall entity that comprises all hives in an area.

The effects are not limited to hives that suffer direct abuse – abuse to the essential, archetypal dimension of *Apis mellifera* will in time harm all bees.

Colony collapse calls for a new way of understanding wholeness: a scientific quantum leap that includes the essentials and is able to read the pictures that nature itself provides for her plight. By ignoring the bee in all bees, we can only harm nature.

To blame beekeepers misses the point. They love their hives and their work. Their rationale is sound by conventional standards. They are victims of the same paradigm we all suffer from: the bee crisis is a crisis of the global mind and we all are responsible. Today, we know more about bees than any generation before us, yet we understand less about them than ever. As our eyes have been opened to detail, our minds have been blinded to the whole. We have learned to decode the genome, but neglected to heed the obvious.

The balance has tipped dangerously so that beekeeping has become bee-losing. We are squandering an irreplaceable key species. Bees are not the first victims of world-wide violence against nature. They are, however, the first species whose loss will be felt by everyone. Apiculture is the stage on which the environmental drama of our time is enacted. We are close to tragedy. And we are all audience, actor and script-writers. We designed the set and chose the theme and wrote the plot. Indeed, we are writing it right now.

6

ENTER: *VARROA DESTRUCTOR*

And so it is that beekeepers can indeed be very happy with all the progress that beekeeping has recently experienced in such a short time, but this happiness will barely continue for one hundred years.

Rudolf Steiner, *Bees*

Varroa destructor is a small, brown, barely visible scaly mite with hairy tentacles, first discovered on the East Asian honeybee *Apis cerana* a hundred years ago. Now it is everywhere. The once harmless parasite has become a global threat. Hive infestation may start with just one female mite attaching herself to the underbelly of a worker bee to suck blood. This way, she enters the hive. Once she has sated her appetite she makes for the brood cell. Each bee larva has its own cell. The mite makes herself at home in the liquid food of the larva and sinks into a stupor. When the cell has been capped (sealed with wax) by nurse-bees and the food eaten by the bee larva, the mite wakes from her honey-trance and attaches herself to the pale body of her embryonic host.

She pierces the helpless larva and feeds on its fresh blood. Grown fat, she deposits her own brood inside the waxen cell. The mites hatch and crawl onto the bee pupa. It is shocking to see magnified images of these dark spots clawing into the embryonic bee, siphoning life-blood.

The mite's first hatchling is usually male. The male waits for his sisters to mature, then mates with them inside the cell. Soon thereafter the male mite dies in the very cradle in which he was born, while his sisters continue to feast off the young bee inside the cell until the bee is full-grown. The sisters leave the cell when the full-grown bee emerges, find another cell within the hive and repeat their cycle. Bees that hatch from mite-infested cells are weaker than their sisters. As the infestation grows in the hive, two or more mites may make their abode in a single cell, leaving the bee crippled or dead.

The mite is prolific in her reproductive vigour, far more than the bee. Mites reproduce in a ten-day cycle, half the time worker bees take to mature. Even a single mite is potentially fatal to the hive. The one intruder soon becomes many, and before long their number is legion. At this point the colony is weakened to the point of collapse. Viruses and bacteria join in the decaying feast and the once-thriving hive is suddenly empty.

The bees are utterly defenceless. Nature, it seems, has contrived an invulnerable predator that grows at twice the speed of its prey: the aggressor's strength grows proportionally to the victim's weakness. Bees seemingly have no chance and beekeepers no solution. Their only hope is a kind of chemical warfare to support the invaded hive. But to attack mites without harming bees, to poison the intruder without spoiling the honey, is difficult.

The culprit, at least, has been found. The mites have been caught in the act. There is overwhelming evidence: the worldwide crime against bee brood is documented in every detail, facet and fact. The only question is how to proceed with the execution of the invader without bees being harmed by the effect. But there is one problem, one fact that has been overlooked: the mites have what in the law courts of nature would amount to a strong defence: they can claim long-lasting coexistence with Asian bees, and in all that time were no less parasitic than they are now. Further investigation reveals that the *Varroa* mites are as ancient as bees. Both belong to the phylum Arthropoda, and both originate in times that predate man, mammal, birds, reptiles and even fish.

What I have described in terms of a cruel attack on entirely helpless bees is in fact a natural process for Asian bees, *Apis cerana*. The mite can only propagate inside the larvae of bees. It has done this for aeons without harm to its host. Something, certainly, has changed, but it's not the mites. The small creatures that were virtually unknown at the beginning of the twentieth century and have since grown to fame as terrorists have remained the same. True, they are new to the West, which they entered via the worldwide bee trade. But what has made these Eastern mites a menace in the West? (A few sturdy hives do survive *Varroa* infestation, but most perish.) Has *Apis mellifera* been unable to adapt or has its health been compromised beyond the point that it can defend itself?

What forced the mites to become suicide attackers? For suicide it certainly is. What kills bees will kill mites in turn. When there are no bees there will be no more bee brood, and no bee brood means no more *Varroa* propagation. The mites are definitely sawing the branch they are sitting on. As are we.

7

MITES AND THEIR MISSION

Nature goes her own way, and all that to us seems an exception is really
according to order.

Johann Wolfgang von Goethe, *Conversations with Goethe*

Parasites are part of nature, of the totality of life. To call them parasites
betrays prejudice. Only a narrow perspective will see them as nuisance or
threat. In nature there is continuous communication between predator
and prey, a subtle co-operation and lawful interaction that escapes the
coarse net of the intellect. What at first looks like competition reveals
itself to be a balanced collaboration.

In his bee lectures Steiner offers the example of the Ichneumon fly,
a kind of wasp that deposits its brood into a living caterpillar. Inside
their slow and sluggish host the wasp eggs mature. When the larvae
emerge from the eggs they make a meal of their host. But in spite of their
ferocious appetite they harm no vital organ. They feast only on those
parts of the caterpillar that it can live without. The caterpillar survives
and the larvae are well fed. Eventually the whole brood will crawl out of
their exhausted prey and take to the air. Fully matured, the females will
find another caterpillar, lay their eggs and start the cycle again.

We call these wasps parasites. They avail themselves of the caterpillar
but never overstep the line that endangers their host. In their own way
they do what the *Varroa* mite has always done: co-operate in order

to exist. Nothing in the life of the mite has changed for millennia, but the same cannot be said for the life of bees. As we have seen, this has dramatically altered in recent centuries, and in the last century in particular. The self-expression of *Apis mellifera* has been increasingly compromised, the essential bee all but extinguished in the constant tearing of the connective tissue between queen and worker. It is this, I believe, that makes the European bee so vulnerable to *Varroa* mite. (Asian bees co-evolved with the mites.)

The absence of connective tissue creates a void which, like a vacuum, invites invaders. Thus the void is occupied by parasites who proliferate beyond control, there to bring an end to what is already dying. The true cause is compromised immunity: compromised by the sum total of interferences and finally destroyed by the manipulation of the queen bee. *Varroa* infestation is only one of many ways for bees to die. Tracheal mites can affect bees in similar ways. The bee disease foulbrood can cause colonies to collapse. To the imagination the reason for *Varroa* prominence is obvious, for the kind of death these mites inflict is a telling reflection of the abuse bees suffer from modern beekeeping practices. The two have much in common.

Mites insert themselves into the reproductive process of the hive and exploit it beyond repair. They intervene between the reproduction of the matriarch and the end product of the bee. Availing themselves of the reproductive process of the hive, mites multiply to the point where it becomes fatal for themselves. Beekeepers do similarly when they manipulate the reproductive process of the hive to multiply their yield. They too have overstepped a mark. Like the mites, they are ruining the very resource on which they depend.

Similar indeed – but not the same. Beekeeper and mite relate to one another like beginning and end, cause and effect. The mites tear the visible tissue; modern beekeeping methods tear the invisible bond. Mites mirror our handling of bees on an organic level. They are executors of processes we began. The *Varroa* plague does not create the problem, but it does reveal it. This becomes even more apparent when we compare the reproductive process superimposed on bees by their keepers with the one associated with the mites. The incestuous *Varroa* inserts her own genetic monoculture into the genetically diverse reproductive process of the hive. Modern beekeeping practices do the same when they force man-made genetic monoculture onto the hive. Both exploit the bee by curbing natural reproductive habits.

The paradigm behind beekeeping will label the *Varroa* a terrorist and so divert responsibility for its own doings. Giving the *Varroa* the fearsome secondary name of 'destructor' testifies to this attitude. From the point of view of comprehensive ecology, though, the mite is a messenger sent to prevent greater harm. I would rename it *Varroa instructor*. The intellectual mindset will battle the mite with the same practices that caused its spread. A comprehensive ecology will learn the lesson that it has to teach.

We have a choice. We can continue to examine the details or we can look to the whole; employ analysis or use synthesis; invest all trust in the reductionist intellect or apply imagination to comprehend the greater picture. We can relegate responsibility to experts or reinstate our own abilities to see the obvious.

During the past five hundred years science has centred on the development of the analytical mind. The results are our virtuoso mastery of machines and our devastating treatment of the environment. The demise of the bee offers a potent metaphor for this. The intertwined history of the analytic, 'microscopic' mind and beekeeping further illustrates the causes of colony collapse.

8

MICROSCOPE AND MIND

A weak mind is like a microscope, which magnifies trifling things, but cannot receive great ones.

Lord Chesterfield

Microscope, or 'small-look at', was the name given to Galileo's early version of the instrument. It is not known who invented the microscope and its twin, the telescope. Legend has it that the principle underlying both was accidentally discovered by children playing in a lens-maker's shop. This is certainly true metaphorically, for microscope and telescope are both children of their time, born at the moment the medieval orientation towards universals gave way to scientific preoccupation with the particular. The births of science and microscope coincide.

Unbeknown to those involved in the pioneering of microscopes, a threshold was crossed. The visual skin of the world was torn and parts began to dominate what was formerly whole. A door opened onto ever-smaller rooms with ever-smaller doors, leading into a maze of detail.

We are too used to this labyrinth to appreciate the inner upheaval that accompanied the arrival of the microscope, but I can still remember the strange mixture of shock, disgust and fascination when I first saw a human eyelash magnified into a thick, unappealing stem surrounded by fat mites feasting on bits of facial debris. The world is never quite the same after such a sight. It changes the nature of reality and leaves a scratch on the surface of what was formerly whole. Several hundred years ago such experiences must have been profoundly disturbing. The plate by Francesco Stelluti (Figure 4) is an early example of an illustration made with the help of a microscope.

The drawing by Stelluti is beautifully executed. The details are accurate. Whoever saw it at the time saw more of the bee than anyone had seen before, but also less of it. For every detail gained there is a loss in wholeness. While the gain is obvious the loss reveals itself only to closer observation. The reason we cannot see this loss immediately is simply that we have become habituated to the microscope in our own mind. Focus on fragments and magnification of detail have become second nature to us. We may be startled by mites clinging to the base of our eyelashes, but see nothing extraordinary in Stelluti's depiction of the bee.

If, however, we look with artistic sensitivity we may notice that the bee has been taken apart and distributed all over the page. Two legs, disproportionately enlarged and turned upside down, are put into the lower left-hand corner of the drawing. Parts of the head, mouth, tongue and a single antenna are neatly fitted into the rectangular frame. There is no lack of rationale in the use of space. All the parts are clearly described and numbered. It is a perfect study. But what seems so well done and beautifully arranged, what may even be viewed with a certain 'once upon a time' sentimentality by a contemporary observer, is simultaneously disturbing to a finer awareness.

A sensitive observer may even feel a touch of the horror the public once felt at the dissection of corpses that became widespread at about the same time. Cutting into a body ruptured more than the skin of a corpse; it violated a taboo, crossed an emotional threshold. The microscope did the same. As soon as it became available there was no limit to dissection. What the microscope cannot see, the lens of the mind can. The scalpel of the intellect will cut further and deeper, and eventually split the atom and decode the genome. The potential of science to take apart and distort announces itself in Stelutti's drawing. This study of a

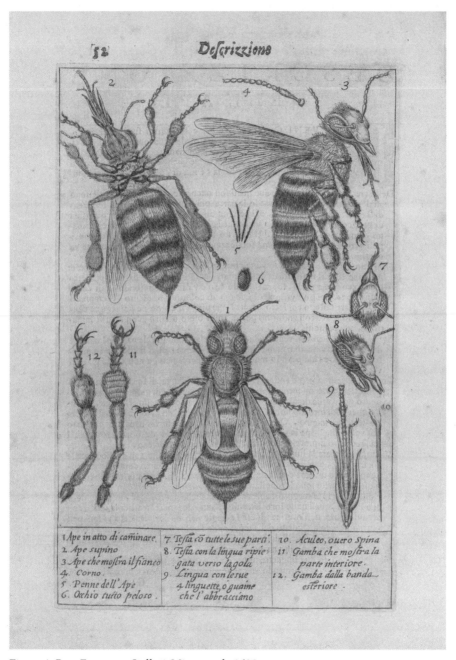

Figure 4. *Bees,* Francesco Stelluti. Micrograph, 1630.

bee is simultaneously an illustration of the dangers of one-sided science. It is a telling revelation of the analytic mind and its opposition to nature: of all that we have done to her and all that we are doing to the bees right now.

Stelluti's micrograph is not just a random drawing found in a manuscript; it is also a scientific icon, a historic document of the first order: the first illustration ever made with the help of the microscope.[1] Nor is Stelluti just any early illustrator. He was one of the most prominent scientists of his time and one of the founders of the Lincean Academy in 1603, then the only academic research institution in Europe. The institute would soon include in its ranks the illustrious Galileo Gallilei, who successfully improved both microscope and telescope. The academy's chosen emblem was the sharp-eyed lynx, and its mission to 'take care of small things if you want to obtain the greatest results' was pursued with telescope and microscope (even the names of the two instruments were conceived by members of the Lincean academy). Galileo focused on what was small by dint of its distance and Stelluti on what was minute but intricate. Insects came under scrutiny, and the first to be examined was the bee.

It is significant that the bee should be the first victim of microscopic dissection. What may seem merely accidental to the analytic intellect is anything but surprising to the imagination. The creature that exemplifies relatedness and the instrument of isolation are natural opposites and, like all opposites, attract each other. The result is the intertwined history of microscope and bee.

Flemish writer Maurice Maeterlinck intuits this relationship in his 1901 classic *The Life of the Bees*:

The real history of the bee begins in the seventeenth century
with the discovery of the great Dutch savant Swammerdam…
Swammerdam founded the methods of scientific investigation:
he invented the microscope, contrived injections to ward off decay
and was the first to dissect bees…[2]

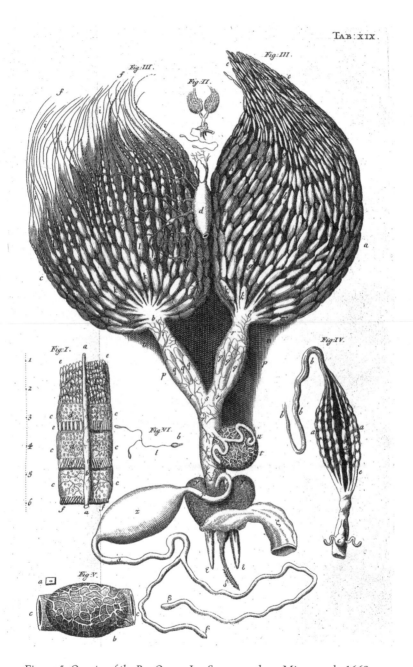

Figure 5. *Ovaries of the Bee Queen*, Jan Swammerdam. Micrograph, 1669.

That Jan Swammerdam invented the microscope may not be true, but that he put it to effective use certainly is. He was expert in the dissection and description of wasps, ants, dragonflies, butterflies and, most notably, bees. In his dissections he found the ovaries and oviduct and surprised his contemporaries with the discovery that the vast, complex civilisation of the bees was a matriarchy, ruled by a queen rather than a king.

It is astonishing to see the rapid progress the microscope allowed in the course of only a few decades. Stelluti's 1630 plate represents a magnification of external parts. Swammerdam's drawings published forty years later expose the insides. Figure 5 shows his famous illustration of the reproductive organs of the queen.

Many of Swammerdam's micrographs are still in use and testify to his skill. Even nowadays these drawings appear awesome and strange, fascinating and repulsive. They have opened a door to an interior world. They expose intimate knowledge of details whose proper application requires a corresponding intimacy with the bee in all bees. Leonardo da Vinci could draw a human womb with the unaided eye. Swammerdam required the microscope to expose the ovaries of the queen bee.

I am in no way opposed to microscopes. They are brilliant and indispensable tools of modern science. I am concerned about the microscope as the external manifestation of a one-sided focus on parts. The instrument that could be an inestimable blessing in the service of a macroscopic ecology turns into its opposite when used in a one-sided manner. It is this one-sided focus that has turned queen bees into exploited tools.

9

MATING MACHINERY

Louis XIV was made for a brilliant Court. In the midst of other men, his figure, his courage, his grace, his beauty, his grand mien, even the tone of his voice and the majestic and natural charm of all his person, distinguished him till his death as a true King Bee.

Louis de Rouvroy, Duke de Saint-Simon,
Memoirs of Louis XIV and His Court and of the Regency

Swammerdam died in 1680. The Duke of Saint-Simon lived after the great Flemish naturalist, yet he still compares his monarch to a bee king, ruling the hive. To the French royalist, a perfect civilisation such as that of bees could only be ruled by a king. The truth of the queen bee was obviously challenging and took a while to infiltrate. Major paradigm shifts take time, and the beginning of the modern era was rife with shake-ups: the earth had been taken off its pedestal by the telescope and the rule of bee kings dethroned by the microscope. The same science that put the sun in the centre also instated the queen. A closer look, however, reveals that the real ruler is neither the sun at the centre of the universe nor the bee queen in the midst of her hive, but science itself. It is those who make the rulers who truly govern.

This is immediately apparent in the case of bees. As soon as the gender of the queen was fixed she became an object of scientific inquiry. Inquiry led to knowledge, and knowledge to total subjugation and abuse. Since then the queen has been a puppet ruler. Behind her stands a beekeeper, and behind the beekeeper stands reductionist science – the real bee king. What science discovers governs the queen and, via the queen, the hive.

After Swammerdam's revelation that the bee king is really a queen, research turned towards understanding the queen's reproduction. What can clearly be seen in fish and birds and rabbits remained hidden from observation in the case of bees. To this day nobody has witnessed the mating ritual of a queen in nature.

The early eighteenth-century polymath Réaumur was one of the first to investigate the bridal mysteries of the queen. A typical exponent of the French Enlightenment, he shed light on the life of bees by introducing glass hives. Réaumur realised that each hive is ruled by one queen only, *la mere de toute people* – the mother of all her people – and that the drones' function was the fertilisation of the matriarch. Prying into the private life of bees, he observed the queen laying her eggs into uncapped cells. He tried and failed to mate queen and drone by confining them in glass containers.

The bridal mystery of the queen remained unsolved until the time of François Huber, the father of modern apiary science, whose research spans the second half of the eighteenth century (see Figure 6). Though blind, Huber saw more deeply into the life of bees than anyone else. He accomplished this extraordinary feat with the help of his wife, Marie-Aimée Lullin, and his devoted secretary, François Burnens. It seems apt that the secrets of the communal bee should be explored in community and through the same careful division of labour that marks the hive.

To aid his research Huber devised a segmented hive that was made of wood and opened like the pages of a book (see Figure 7, overleaf). Segmentation allowed him to take the hive apart and examine the interior of the colony. Swammerdam took bees apart, but Huber dissected the hive. Where hive dissection did not suffice he applied the microscope in the mind. Seeing through the lens of ingenious experiment, he discovered that the queen was fertilised by drones outside the hive.

Huber was the first scientist to focus exclusively on bees. He was

Figure 6. Swiss naturalist *François Huber* (1750–1831).

also the first to attempt artificial insemination of a queen, using drone liquid on the tip of a hair pencil (an fine camel-hair brush), but failed to produce results. Still in its cradle, modern apiary science attempted to achieve with artificial means what nature had hidden away. The clash between bee science and bee care becomes apparent.

Further progress in understanding bee reproduction was made by Johann Dzierzon in Silesia in 1835. Dzierzon was a pioneering apiarist and a Catholic priest. He built the first movable-comb hives to harvest honey without disturbing the hive. His practical invention also allowed him to better observe bees. Studying eggs of fertilised and unfertilised queens, he discovered parthenogenesis, the ability of the bee queen to reproduce without male fertilisation. Such virgin birth, however, produces only male drones. This means that drones are produced from unfertilised eggs and workers from fertilised eggs. When he published his discoveries in 1845 he met with strong resistance from both the scientific establishment and the Catholic Church. Production of male offspring without male participation was anathema to the science of the time, and patriarchal sensitivities added to the force of the reaction. The Church condemned Dzierzon's findings as heretical. Catholic dogma

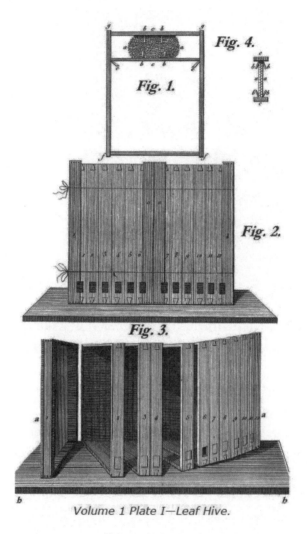

Volume 1 Plate I—Leaf Hive.

Figure 7. *Huber Hive (Leaf Hive)*, François Huber, 1806.

reserved virgin birth for Mary, queen of heaven, and would not grant it to the queen of bees. Eleven years later, however, Dzierzon's observations were confirmed through the microscopic studies of the German zoologist Karl von Siebold. Scalpel and magnification proved Dzierzon's conjecture and catapulted his theories into public awareness.

Through Dzierzon's research the reproductive biology of the queen bee became sufficiently understood to make manipulation possible.

Until this point in time beekeepers had relied on the natural reproduction cycle of hives, where swarming increased stock. Dzierzon's insights allowed the creation of artificial swarms: separating a sufficient portion of bees from the mother hive and hiving them with an artificially reared queen. This led to a revolution in the cross-breeding of bees and the habit of forcing foreign queens into hives to improve productivity. The biological integrity of the hive was compromised for convenience and profit.

Cross-breeding, of course, was not new. Cattle, sheep, goats, dogs and horses have been cross-bred for ages to secure superior traits or hybrid vigour. Cross-breeding bees, however, is a different matter, for every hive is an entity in itself, a kind of superorganism composed of independent parts. The bee queen does not correspond to a cow or she-goat. She is to the hive what the reproductive system is to a mammal.

What remains external with a warm-blooded animal becomes intense interference with the organisms of the hive: a first attempt to alter forces that in the case of mammals have remained inaccessible to human intervention until recent advances in genetic engineering.

Dzierzon himself crossed Italian bees with local bees, and sent their progeny to many countries in Europe as well as to America, and so began the worldwide bee trade. Around 1900 the understanding of the mating biology of the bee took another leap through the exact anatomical drawings of the queen's reproductive organs by American entomologist and artist Robert E. Snodgrass. *The Anatomy of the Honey Bee* is a classic of microscopic dissection showing longitudinal, median and vertical sections. The alimentary canal, the tracheal, respiratory and nervous systems are separated and artfully illustrated. I am reminded of the segments of Huber's hive now applied to the organism of the bee (see Figure 8, overleaf).

Microscopic dissection, rightly applied, could add much to our understanding of the bee in all bees. But without awareness of its limitations, such knowledge is easily applied to the bees' detriment. Snodgrass's discovery of the valve fold is a telling example. The valve

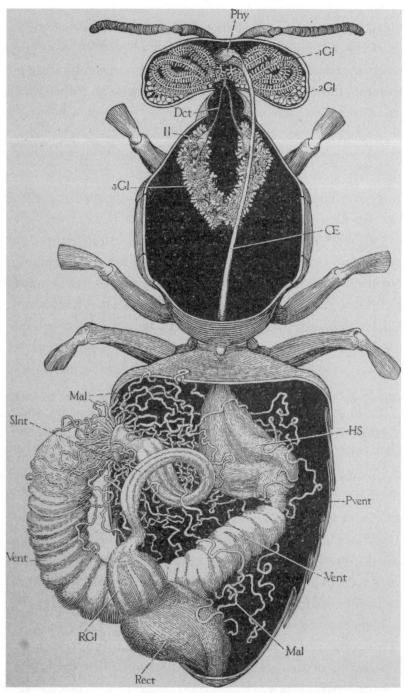

Figure 8. Illustration from *The Anatomy of the Honey Bee*, R.E. Snodgrass, 1910.

fold is a small flap of tissue which prevented artificial insemination of the queen until the 1940s. It was only when Harry Laidlaw at the USDA Baton Rouge laboratory made the link between failed inseminations and the function of the valve fold that success was guaranteed. Circumvention of the fold became the key to fruitful insemination. The rest was a matter of fine-tuning. The method applied nowadays uses a fixed transparent tube to hold the queen securely in place, a ventral hook to open her sting apparatus, and a syringe to inject semen.

Today, a state-of-the-art inseminator includes microscope and computer screen. Mating with the machine has replaced the ritual flight of the queen. Photographic documentation of this procedure is also available on the web and I recommend that readers view the visual evidence. Unless we expose ourselves to reality, reality will not change.

To witness the plight of the queen in this violent, intrusive, mechanical procedure, her long, slim body anaesthetised and held in the vice of technological contraptions, might provide the shock we need. Compare this cold, clinical and sterile procedure with the ecstatic mating flight of a virgin queen, her foray into the open to mate with dozens of drones in mid-air, high-flying suitors from many local hives, each carrying the strength and expertise of their origins imprinted in their genetic material. Compare clinical insemination by way of a syringe with this yearly ritual of drones giving their life for the future of bees. I urge you to immerse yourself in the contrast of such pictures, for unless we apply our imagination and feel appropriate pain and outrage, we ourselves will remain just as anaesthetised as the queen in her tube. This total exploitation is more than a metaphor. What we do to the queen has an immediate impact on the environment where the queen's natural mating ritual serves biodiversity.

The hive is an ecological concentrate, a condensed version of the mutually supportive web that maintains its surroundings. The generous pollination services bees provide to the environment are the very opposite of the genetic monoculture superimposed on inseminated queens. How can we expect bees to contribute to the biodiverse propagation of plants when we superimpose monoculture on their own propagation? Bee propagation is biodiversity in a nutshell. Every hive is an entity and every entity has its own strengths, memories and skills. One hive may give birth to vigorous honey gatherers; another may excel in expert wax makers or particularly fecund queens. Whatever the strength, drones will carry it to their meeting with the queen and so give to the future monarch the gift of their hive. After her maiden flight she will return with semen that represents a cross-section of experience from local hives.

What we term reproduction is a communication between hives, an exchange of experience between local experts in being a bee, and centuries of experience with the flora, weather patterns and seasons and everything else pertaining to the environment. Bees are the ecologists of nature, and the queen's mating flight is their yearly symposium, held in the care of their environment.

To interfere with this meeting disrupts the hive in all hives. We do this when we suppress the mating flight, when we breed and import foreign queens (employing bees in Montana that are experts in the flat Australian bush), and most of all when we artificially inseminate queens. Artificial insemination continues the artificial breeding of bees. It offers more of the same in greater intensity.

The historical quest to understand the queen's reproduction was a first attempt at genetic manipulation and a forerunner of genetic engineering. A beehive is a superorganism and the queen its genetic source. Cross-breeding and, in particular, manipulation of the queen are to the hive what genetic manipulation is to the human body. Both are intimately linked to the use of microscope, dissection and analysis. Dissection is analysis of the body; analysis is dissection in the mind. There is a straight line from Stelluti's plate to today's emerging possibilities of genetic engineering. The mindset that can disproportionately enlarge parts of a bee and put them in arbitrary places is the same that can insert a gene from a tomato into a pig and vice versa. This is the mindset that is capable of enlarging noses, changing the colour of hair, the shape of organs and altering the brain. All these things are or will be soon possible.

There are further parallels between macro- and micro-manipulation. The beginning of genetic research was intimately linked with *Apis mellifera*. Gregor Mendel, the founder of modern genetics, was inspired by Dzierzon's discoveries and began his own genetic research with bees. When he found the busy insects hard to control he changed to peas. His insights into the laws of hereditary are foundational to modern genetics. What Dzierzon discovered with bees became a science in its own right through Mendel just two decades later. A similar time span separates the first successful insemination of the queen and the cracking of the genetic code in the middle of the last century. Bee manipulation is the macro counterpart to genetic manipulation. Both have grown from the same root and will most likely bear the same fruit (The problem of genetic manipulation has been brilliantly explored by American biologist Craig Holdrege in his book *Genetics and the Manipulation of Life.*)

The devastating result of queen breeding and artificial insemination took a hundred years to become obvious. It may take a similarly long time for the drawbacks of genetic manipulation to become apparent. From this perspective the bee crisis may be the canary in the coal mine, a wake-up call before we dig ourselves deeper into the cul-de-sac of genetic manipulation.

Colony collapse teaches us that nature is patient, but not forever; that immediate results are no guarantee of long-term success; that co-operation is the key to sustainability. It can teach us that time is an entity in itself and that the future needs to be taken into account. The bee crisis is a mirror into which we must look deeper in order to understand ourselves and the problems we inflict upon the world.

10

THE LOGIC OF DESTRUCTION

Krishna: I have heard the earth complain.
Yudhishthira: What did she say?
Krishna: She said, 'Men have become arrogant. Every day they wound me anew. There are more and more of them, more violent and driven by thoughts of conquest. Foolish men trample me. I shudder and ask myself: what will they do next?

The *Mahābhārata*

Many beekeepers regard chemical warfare against mites as the only way to stop colony collapse. Artificial breeding of queens is often also seen as part of the solution as it ensures greater control over the genetic material. It is only a small step from queen breeding to the genetic engineering of bees. The temptation to engineer mite-resistant strains is great; to many it will seem the only way forward. A mindset accustomed to seeking solutions in the minutiae of matter will gravitate towards the engineering of a super-bee. There are other solutions, but these cannot be found as long as the intellectual paradigm rules supreme. It is difficult to transcend a way of thinking that permeates our culture, powers our machines and informs technology: it is the mindset taught in every school and reinforced in every university.

Yet transcend it we must. Our destructive logic will otherwise treat bees with the poison that has caused their demise. The only way to stop the momentum is by means of emotional awareness, a presence of mind that is more than mere intellect. Emotional awareness is an act of compassion and a full engagement of all that is human in us.

I experienced the powerful force the intellect exerts upon the mind a couple of years ago when my friend Ann showed me a high school assignment in which students were given an extraordinary hypothetical task. They were to imagine that an atomic war had caused worldwide devastation and that only fourteen people survived the catastrophe. There is only one atomic shelter and it is equipped for only seven survivors. The students must decide whom to admit and whom to abandon. To help them decide, they are given a biographical description of each survivor: one is handicapped, another is a labourer with few skills, one is a nurse, another is a writer and so on.

My friend was outraged at the task. I too experienced feelings of unease and undefined anger. I knew something was horribly wrong. However, had I had the opportunity then to confront the teacher about this assignment, I doubt I would have made any headway. The strength of the arguments at my disposal did not match up to the strength of my feelings. I envisaged such a confrontation and was surprised how easy it would be to find arguments against the position I held. After all, the task was modelled on a situation that could well occur. It could be argued that such an assignment may help prevent catastrophic events. It could not be denied that it would make students think and that it offered ample opportunity for heated discussion. It was much easier to find rational support for the task than arguments against it.

This taught me that it takes time for the head to catch up with the heart. I realised how easy it is to be swept along by the persuasive intellect and the hypnotism of catastrophic events. But it also taught me that it is possible to stop the destructive momentum of the intellect by acknowledging our feelings.

It took me a while to see through the clever construct of the assignment. The exercise asks no less from the students than to make decisions about life and death: they are given licence to kill in the service of humanity.

It forces the student to look at the survivors in a purely rational way, estimate their abstract value and apply it in the name of the survival of the race. The assignment gives permission to murder and at the same time acquits their conscience by virtue of the dictates of catastrophic events.

In fact, the attitude required from the student was the same as one that would have led to atomic war. Whoever pressed that fatal button would have done exactly what the students were asked to do: look at human beings in an abstract way, make rational calculations and estimate the fallout. Most likely the intention would also have been for the greater good, led by similar moral imperatives and rational musts. The button-pusher too would have had to make the difficult decision between the survival of one nation against another. Thus the exercise constructed to deal with future atomic war promoted the very attitudes that would lead to it. I am sure the teacher meant no harm, yet the mindset he or she applied did.

My encounter with the assignment made me aware how easily thought is caught in its own net. To escape its brilliant but destructive reasoning, a new form of cognitive feeling – a knowing-before-the-intellect-takes-control – is needed. Only through compassionate awareness, through experiences that are felt with our whole being, can we hope to escape the vicious circle of self-enforcing arguments. Thought is quick to fall prey to its own momentum. It rushes to conclusions and happily jumps over the obstacles of inner protest. It takes clever shortcuts, and unless we slow our mental operation to the speed of soul, our feelings fall by the wayside.

This same exercise could easily be applied to colony collapse. Students could be given the following assignment: bees are dying worldwide through the spread of a destructive mite. This mite is difficult to combat without harming bees and polluting honey. There are a number of possible solutions. The first is strict chemical control of all hives in apiaries. The second is enforced artificial breeding of bee queens to ensure mite-resistant strains. The third is the genetic engineering of a new mite-resistant bee, a docile, perhaps even stingless and more productive version of *Apis mellifera*. Discuss these possibilities and find the best possible solution for a world in need of pollination.

There are apparently three choices and yet there are none. The mind is trapped within a labyrinth of its own making and appeased with the illusion of choice. To engage at all with the argument is to submit to its suppositions. To think along the lines suggested means we become part of the problem and accomplice to the mindset that created it.

Here, as in the high school assignment, it takes awareness of a particular kind to step out of the head and into the heart. Without engaging emotional and artistic sensibilities it is impossible to withstand the attacks of the intellect. Above all, it takes imagination to see the extent of the implications and act against them.

11

MACROSCOPE

Science has made us gods even before we are worthy of being men.

Jean Rostand, *The Substance of Man*

In earlier days of science, a one-sided focus on parts represented a major advance. It helped to clear the cobwebs of outdated conceptions. Since then, however, sole reliance on the analytic mindset has become counterproductive. A new balance is needed. The microscopic mindset demands a macroscopic complement. For every step we take into detail, a corresponding move ought to be made in the opposite direction. But it is not enough simply to add an understanding of wholeness to detail if the whole is still understood in terms of its parts.

The solution is not a new set of concepts or a different turn of ideas, larger frameworks or improved methods. Paradigm shifts are not enough. They change the products of the mind but not its way of production. What is needed is a thorough paradigm metamorphosis, a change that transforms the way we perceive, investigate, cognise, feel and act. Otherwise the same mindset simply continues under a different guise.

If we do not overcome our 'partiality', nature will fall apart. Respect is lost in detail. It is hard to have an emotional relationship with a chemical formula and there is nothing sacred about a genome. When all regard is lost nature becomes a commodity, the beehive a factory and the bee queen a genetic machine. Nor does it stop at this point. Analysis reduces more than objects. In taking apart, we ourselves become partial. Head and heart separate. Thinking and feeling run on different tracks. This can be clearly seen in beekeepers who form strong emotional attachments to their hives, yet treat their bees with scientific detachment. Like everyone else, they suffer from the intellectual epidemic of our time. The mindset that separates bees into their parts also separates emotion from understanding. Something in us has to be separated, isolated, cut apart, to follow procedures so obviously hostile to bees.

This schizophrenia is the most common ailment of *Homo intellectus* (of the subspecies *destructor*). It is at the core of the ecological crisis and was already apparent in pioneers of apian science. As we saw in Chapter 9, François Huber dedicated his life to the study of bees. His love for his hives was infectious and inspired his secretary and his wife to similar devotion. Their collaboration is one of the most moving accounts in the history of science. Yet in spite of his dedication Huber was the first to attempt artificial insemination of the bee queen. He crossed a territorial taboo for anyone in touch with the essential bee. Knowledge separate from emotional relatedness is prone to abuse. Huber's love for bees clashed with his passion for science. Care and research begin to live side by side, in separate dwellings.

The same tendency can be observed in Jan Dzierzon, the next giant of apian science (see also Chapter 9). The Silesian priest was so passionate about his bees that he often neglected his congregation for his hives, even missing the Mass he was supposed to conduct. Tucked away in his remote parish, Dzierzon wrote:

> In spite of my isolation I am content here because I am always close
> to my beloved bees. Bees transform even a desert into a Paradise for
> anyone who keeps an open mind to the works of the Almighty.

Again we find a devoted, almost religious love for bees coupled with a mindset that could pioneer the breeding and trading of queens, as this quote from *Dzierzon's Rational Bee-keeping* demonstrates:

The objection, that voluntary swarming, as being the natural method of increase, is to be preferred to artificial swarming, is not sound, since we do not keep bees for themselves, that they may follow their own instincts, but for the sake of the profit, that we may gain from them as much honey and wax as possible.[1]

This shows that, in spite of his affection, Dzierzon was willing to compromise the essential bee for the sake of profit. When science and care exist in separate parts of the human being, bees suffer abuse from the very hands that love them most.

This schizophrenia found painterly expression in an iconic artwork *An Experiment on a Bird in the Air Pump* by the English painter Joseph Wright of Derby (see Figure 9, overleaf). The painting dates from 1768, the time of Huber and his experiments, and depicts a 'natural philosopher' – the forerunner of the modern scientist – recreating one of Robert Boyle's air-pump experiments. Boyle was an early experimenter and is nowadays considered a pioneer of scientific method. His famous '*Experiment Number 41*' demonstrates the reliance of living creatures on air: a bird is put into a bell-shaped glass container from which the air is removed by means of a pump. Boyle describes this experiment in 1660:

The Bird for a while appear'd lively enough; but upon a greater Exsuction of the Air, she began manifestly to droop and appear sick, and very soon after was taken with as violent and irregular Convulsions, as are wont to be observ'd in Poultry, when their heads are wrung off: For the Bird threw her self over and over two or three times, and dyed with her Breast upward, her Head downwards, and her Neck awry.[2]

Public demonstrations of the experiments became popular among the educated classes. In his painting, Wright captures the moment the bird begins her death dance. The 'natural philosopher' looks straight at us and so makes us part of the audience. Various characters flank the scientist: two

Figure 9. *An Experiment on a Bird in the Air Pump*, Joseph Wright of Derby. Oil on canvas, 1768.

self-involved lovers uninterested in the procedure, a boy in the
lowering a cage, a man in the foreground looking with undivided atte..
at the experiment, another dispassionately taking the exact time, and a
third addressing one of two girls. The dramatic character of the painting is
enhanced by the single light source that illuminates the scene. Most light
is shed on the anguish of the two girls, whose agony is the major theme of
Wright's nocturne. The painting captures powerfully the contrast between
intellect and compassion, science and emotion. The two girls stand for the
compassionate part in us all that cannot bear what it sees yet has no means
of changing it. Like the bird, they are victims of a science without soul.

The scene reminds me of Rupert Sheldrake's introduction to his book
Dogs That Know When Their Owners Are Coming Home, in which he recalls
his early fascination with animals. Sheldrake was particularly taken with
the homing instinct of pigeons. When his passion turned profession, a
gulf opened between his sentiments and his professional reality. 'Studying
a living organism was to kill and cut them up,' he writes. Vivisection
followed dissection when he worked in a pharmaceutical laboratory.
Injecting chemicals into tortured rats, mice and guinea pigs to test
painkillers was far from his initial intent, and he found himself in much the
same situation as the girls in Wright's painting. Eventually his own homing
instinct led him to develop a more compassionate scientific approach.

Colony collapse is a calling for a global homing instinct towards
compassion. It is an urgent indicator that the separation of science and care
must not go on. This separation has had its time and fulfilled its mission.
It has taught us to become objective through detachment. Now we must
learn to remain objective in relatedness.

In quantum physics, science itself has already met the inseparable unity
of mind and matter, subject and object. What was discovered on the micro-
level needs to be applied to the macro-reality at hand. The end result of
traditional science must become the starting point of a new compassionate
ecology. This requires a radical reorientation and the development of new
capacities to counterbalance microscopic trends. I call these capacities
macroscopic. They are inner tools able to comprehend wholeness just

as the microscope comprehends parts. These are means that link us emotionally and so bridge the intellectual gap that allows abuse: methods of inquiry in which compassion is not a bystander but an active participant. Such methods already exist.[3] They are part of a newly emerging scientific trend that uses imagination and compassionate awareness in the same objective ways a scientist uses the intellect. It took two and a half thousand years to develop the intellect into an exacting tool. It will also take time for compassion to achieve an equivalent level of maturity.

Steiner was one who worked in the tradition of compassionate ecology (compassion not as it appears in ordinary life, but systematically developed as a cognitive capacity). His ability to foresee the bee crisis is testament to the effectiveness of methods that stand to become the foundation of the new paradigm. To understand these methods we had best start where Steiner did: with the German literary giant Johann Wolfgang Goethe.

12

GOETHE: THE APPRENTICE OF NATURE

Goethe is the Kepler and Copernicus of the organic world.

Rudolf Steiner, *Goethean Science*

Goethe is one of the giants of world literature (see Figure 10, overleaf). He is to Germany what Shakespeare is to England, and Dante to Italy. His importance, however, is not exhausted by his poetic achievements. He was a man of universal interests, and his passion for science led him along innovative paths that remain unexplored by conventional research. He wrote a major work on colour theory and contributed to geology, botany and animal morphology (a term he coined). Goethe's colour studies led to the discovery of the magenta spectrum, and his morphological insights proved the existence of the *os intermaxillare* (a bone of contention in eighteenth-century anatomy).[1] Yet his significance lies not in his single discoveries, but in the method he employed to reach them. It is this method that made Steiner hail him the 'Kepler and Copernicus of the organic world'.[2]

By Goethe's time science already ruled supreme. Copernicus, Galileo, Kepler and Newton had brought it to centre stage. The new mindset had taken root, silencing more cautious voices, such as the English and German Romantics and the American Transcendentalists. Whoever studies the work of Blake, Keats, Wordsworth, Coleridge, Schiller, Novalis, Emerson and Thoreau will find the seeds of Goethe's approach. His genius was to

Figure 10. *Johann Wolfgang von Goethe* (1749–1832).

systematically pursue what was only touched upon by his contemporaries.

Goethe's work with plants exemplifies his innovative approach. In the eighteenth and early nineteenth centuries, botany was popular among laymen and scientists. Carl Linnaeus was renowned for his success at systematising plants and animals. Bringing order to the vast variety of forms, he became the father of modern taxonomy, and his distinction into phylum, genera, species and so forth is still in use. He baptised in Latin and is responsible for the rather hopeful term *Homo sapiens* (wise human), as well as *Apis mellifera*.

Linnaeus achieved his order by detouring through parts. Comparing leaves and petals, stamen, pistil, ovary and fruit he was able to systematise the bewildering diversity of plants. To establish the degree of relatedness, he concentrated on what distinguished one plant from another. Goethe studied the work of Linnaeus carefully and derived much benefit from it. Yet he remained dissatisfied with the result. What was put together after first being taken apart lacked life. He expressed his reservation in his dramatic masterpiece *Faust* where Mephistopheles explains the nature of science to an unsuspecting student:

He who would know and treat of aught alive,
Seeks first the living spirit thence to drive:
Then are the lifeless fragments in his hand,
There only fails, alas the spirit-band ... [3]

The 'spirit-band' alludes to the life lost in the reductionist approach prevalent in his time. The clarity gained in detail is bought at the cost of wholeness. It was this approach that Goethe sought to complement with his own work. Looking for the life lost in analyses and the wholeness obscured by reduction, he developed a radical phenomenology: radical compared with conventional approaches and tender in relation to nature. Francis Bacon's famous quote that 'Nature needs to be put on the torturing rack to wrest her secrets from her' is the very opposite of Goethe's approach.[4] To Goethe, nature is intent on revelation and wishes to disclose her secrets.

Conventional science, with its Baconian undertones, proceeds by first framing a hypothesis, then deducing consequences and, finally, testing these consequences by means of experiments whose results confirm or contradict the hypothesis. In essence this method is a court procedure (Bacon was a lawyer) and is inherently violent when applied to living phenomena. Nature gets a hearing before the jury of the analytical mind. Accused of hiding vital secrets, it is cross-examined. If results do not match expectations nature is further taken apart.

To use such a method is like trying to read a letter by analysing the ink. The answer will not be wrong. The chemical composition of ink may be quite accurate. But it is not what the letter was about.

Goethe, suspicious of the science of his time rather than nature, reversed Bacon's approach. His method resembles a courtship rather than a court case. He patiently observed and held back his judgement until the observations gave their own verdict. He aimed to be an instrument through which nature arrives at her own conclusions about herself.

Where Linnaeus looked for distinguishing detail, Goethe sought the uniting principle that informed all parts. He realised that the plant we see is only part of the real plant, a temporal manifestation of the whole that only reveals its totality over time. To align himself with the living principle he studied the plant in its totality: he followed the plant-growth from seed to cotyledon to leaves spiralling around the stem; he proceeded with the formation of the compact bud and the first tender opening of the blossom, the full bloom surrounding pistil and stamen, the formation of pollen, the slow ripening of fruit and finally back to the seed again. He broke completely new ground when he linked successive stages with his imagination and so made inwardly whole that which was outwardly separated. Goethe made the plant grow in his mind as it would in reality. This is by no means easy. To transform our inner life of pictures into an exacting scientific tool takes disciplined practice, a careful tending of capacities rarely used, a major shift in inner orientation. A taste of his method, however, is well within our reach.

The following exercise is relatively easy and may serve as a first if modest introduction. Take a plant, for example a daisy. Find a bud that is not yet open and observe it carefully. Take time to imprint its shape, colour and overall appearance in your mind. Then close your eyes and see if you can clearly recall it. If this proves difficult I recommend making a coloured pencil sketch, then trying again. Now find a bud that has already opened a little and proceed as before.

Once you have familiarised yourself with both stages, picture one after the other in your mind's eye several times. See the first as clearly and with as much detail as possible and then replace it with the second. When I conduct this exercise with students, most have the experience that the two separated pictures do not remain isolated. The imagination fills the gap and opens bud into blossom. It does in us what the flower will do in nature. It links separate parts through movement. Isolated instances join to become a whole. We approach the life of a plant through what is alive in us and so enter the dimension of time.

Goethe pursued such exercises systematically and over long periods. Training his imagination until it was ready to produce the whole plant, he entered the time body and became intimate with organic growth. He began to understand the stages of plant development as an expression of the dynamic whole. It was by applying this method to the human body that he proved the existence of the *intermaxillare* bone.

More is made whole in this process than the bringing together of parts. Matching a process in nature with a process in himself, Goethe bridged a yet more important gap: the ever-widening abyss between self and world, observer and observed, subject and object. This distance is the core problem of ecology, and more important than the bee crisis, global warming, deforestation, water scarcity or any other environmental issue. I say more important because it addresses the central cause of all others.

This separation from nature makes us strangers in our own world. It produces an endless accumulation of unrelated facts that makes the planet come apart. It is this separation, this constant, painful tearing apart that makes us able to leave our footprints on the moon while we lose the earth beneath our feet. Because of this separation we discover the interdependence of observer and observed in micromatter and yet miss our relatedness to nature close at hand: the ability of our imagination to compassionately relate and so close the gap that the intellect has opened.

Only such intimate knowing can overcome scientific separation and its consequences. The intellect that reduces the plant reduces my relationship to it, and does the same to bees, cows, rivers, landscapes, the whole of nature. The result is diminished care and, eventually, abuse. The imagination, on the other hand, makes whole because it is whole. It links and unites us with whatever we investigate. The results are relatedness, care and collaboration.

In light of the present environmental crisis Goethe's approach offers a much needed complement to conventional science (a theme comprehensively explored by British physicist Henry Bortoft in *The Wholeness of Nature*[5]). Goethe's approach opens the possibility of a new science of relatedness, of empathic knowing, of the compassionate ecology that the bee crisis demands.

13

COMPASSIONATE ECOLOGY

We must trust the perfection of the creation so far, as to believe that whatever curiosity the order of things has awakened in our minds, the order of things can satisfy.

Ralph Waldo Emerson, *Nature*

To fully appreciate Goethe's contribution it is important to recognise that science itself had humble beginnings. Seven or eight centuries BC, the mythological mind ruled supreme. It was as persuasive a paradigm as the scientific worldview is now.

Pythagoras was the first to initiate the rudiments of scientific comprehension. It might not have meant much to many of his contemporaries when the Greek sage related musical intervals to mathematical proportions. It may not seem much to us either when compared with the sophistication of quantum physics and the elaborate technologies employed in laptops. Yet it signified an extraordinary breakthrough, a gigantic leap into a totally new way of relating to the world. Understanding nature by means of thought laid the foundations for Western science, and all that we know of science, laptops and quantum physics has followed in its wake. Yet in Pythagoras's time such insights were as new and ephemeral to the Greeks as Goethe's approach is to our age.

Like many innovators, Pythagoras was persecuted and his school attacked. Major paradigm shifts always upset the status quo. Giordano Bruno was burned at the stake for thinking twentieth-century thoughts in a sixteenth-century environment, and Galileo was arrested for supporting Copernican views. Goethe, living in more civil times, was celebrated as a poet but marginalised as a scientist. His discoveries were acknowledged, but his participatory science was largely ignored.

There were notable exceptions such as Alexander von Humboldt, the great scientist-explorer who laid the foundations for physical geography and meteorology, and successfully combined his own methods with Goethe's approach. In a letter Humboldt writes, 'elated through Goethe's approach to nature, I was equipped with new organs of perception.'[1]

The nineteenth-century German naturalist Carl Gustav Carus developed the concept of the vertebra archetype by following Goethean lines of inquiry. Other scientists who applied Goethe's method were the influential nineteenth-century Norwegian natural philosopher Henrik Steffens and the German Lorenz Oken. Yet their efforts remained isolated attempts. The time had not yet come. Science, seemingly victorious on all fronts, saw no reason to examine its methods.

In light of bee colony collapse and global environmental crisis, Goethe's method warrants another look. So does the work of those who followed in his wake, among them Rudolf Steiner, who began his career as a Goethe scholar. Steiner took what Goethe had begun and developed it a good deal further. He engaged further capacities, among them the power of clarified, transformed feelings or, as I call it, compassion: compassion not as it appears in ordinary life, but systematically developed as a cognitive force by dint of will.

To the conventional scientist this might sound naive, sentimental, perhaps even ridiculous. When we think of compassion we think of feelings, and these are not something that can be trusted when it comes to science. Here we need to remind ourselves that at the dawn of science the emerging intellect was just as unreliable as our emotions are now. A brief study of the history of science and the flights of fancy taken by some of its most illustrious proponents should convince anyone. What we finally meet in our textbooks is a knowledge that has been carefully distilled over centuries. Clarity of thought developed with discipline over time. Clarity of emotion could do the same.

The difference between thought and feelings is that our thoughts keep us detached, while our feelings make us related. The bee crisis shows how detachment can become destructive and that a new approach is needed, one that is related as well as objective. Here Steiner has pioneered methods that could become fruitful if only we can leap the hurdle of prejudice. While I believe that acceptance of any or all of his esoteric research is a personal matter, his method, I argue, is indispensable for a compassionate ecology. Steiner's approach is demanding. Here, as in the case of Goethe's work, a preliminary example may help to point in the right direction.

To read about the bee crisis may or may not bother us. Mere thought often leaves us detached. If, however, we see an image such as the anaesthetised queen bee held in a test tube with a syringe stuck into her abdomen, we become affected. Engagement increases the moment we vividly produce this picture in our mind and elaborate on it with imagination. The more accurate and lifelike these pictures are, the more painful they become. Through imagination we begin to resonate with the queen and with the abuse she suffers. We begin to feel the cruelty of the procedure, the outrageous imposition of modern technology on bees, the total lack of compassion and care.

The experience can be increased if we contrast this cold clinical procedure with the euphoric mating flight of the queen. Using our imagination we can indeed see what nobody has seen before: the queen's ecstatic rise into the light-filled open air to mate amid the great humming congregation of drones ready to convey the genetic wisdom of generations to the new mother of the hive. We can transport ourselves into the sun-flooded spring rite in which the whole environment secures its productivity by means of the mating queen.

Immersing ourselves in such pictures, we not only feel, we know.

Empathetic imagination becomes a cognitive tool. We begin to understand what experts are apt to miss: the level of abuse inflicted on bees and the real reason behind colony collapse.

To know through empathetic imagination what experts habitually overlook is crucial in times when scientific research dictates public opinion. In regard to bees, cows, fish, rivers, seas, forests and climate we must all develop ecomorality and not depend on the dictates of the paradigm that endangers these. The imagination is a first step in this direction. It allows us to see beyond abstraction, feel abuse when it occurs, resist exploitation and act against injustices inflicted upon the environment. The imagination offers the macroscopic perspective that alone can help an otherwise destructive science to serve the whole.

This initial compass of compassion helps us to navigate the labyrinth of scientific fact and steer clear of abuse. Yet it is still far from the exacting tool Steiner used when he predicted colony collapse on the basis of queen breeding. It is, however, related to it in the same way as the humble beginnings of science are to its later accomplishments. If I understand Steiner correctly, he further developed the method of imaginal participation described above through scientific accuracy. He intensified Goethe's method by combining scientific thought with compassionate feeling through enhanced inner activity. This approach overcomes detachment by engaging the whole human being. More of us is called forth to know more of the world: the result is the objective relatedness and increasing acquaintance with the essentials. This is the only antidote to abuse in times of powerful but partial knowledge. This intimate knowing not only helps us avoid abuse, but allows new insights into the essential bee, cow and cod. From such insights come new ways of sensitive collaboration.

The approach Steiner suggests – the application of compassion to help overcome the gap between observer and observed – is by no means reserved to natural phenomena such as plants or bees. It is equally applicable to economy, sociology, psychology, philosophy, history – indeed, to any phenomenon. Even time can be viewed in a participatory way, something we might call 'chronoliteracy'.[2]

All this is by no means easy. It takes a great deal of dedicated practice to progress in this direction, and much in the present paradigm militates against such work. To combine science with compassion, clear thought with clarified feeling, requires effort and a thorough transformation of our mental make-up. Pursued in the right way, knowing becomes an affair of the whole human being. The entire psychophysical organisation becomes active in the understanding of nature, that nature to which we ultimately belong and with which we have coevolved. When the totality of our capacities is employed to grasp the totality of nature, ecology becomes complete. The result is compassionate care and conscientious collaboration.

14

FROM CONSCIENCE TO COMPASSION

Our mind is capable of passing beyond the dividing line we have drawn
for it. Beyond the pairs of opposites of which the world consists, other,
new insights begin.

Hermann Hesse

Compassionate ecology is a powerful remedy for the scientific
detachment at the core of the environmental crisis. The separation of
intellect and emotion is not only symptomatic for scientists, but for
everyone living within contemporary culture: we all have our share in
the schizophrenia of our time.

Alienation through detachment goes back to the beginnings of
science. But while detachment was a developmental necessity, it must
not remain the sole determinant of the future. History itself points to the
solution. For, looked at more closely, intellect and emotion, detachment
and compassion are not as separate as they appear at first sight. In fact
they belong together, as twins born at the same time, though in different
places. At the very time that Pythagoras initiated Western culture into
science, Buddha was developing universal compassion in the East.
The Greek sage and the Indian saint are exact contemporaries, their
births but ten years apart. Both lived in a time when newly emerging
intellectuality replaced mythical conceptions of the world.

Pythagoras brought clarity to the mind. Buddha brought objectivity to the soul. One developed external science, the other compassionate knowing and moral capabilities. One gave the impetus to master the outer, the other the inner world. Scientific thinking did not exist before the sixth century BC. Egyptian and Babylonian conceptions were still imaginal and allied to religious cults. (By imaginal I mean the collective pictorial thinking of the pre-intellectual past, as well as its modern metamorphoses in the work of Goethe, Novalis, Keats and Steiner.) Pythagoras' new impulse toward science in the West mirrors that of Buddha's complementary intentions in the East. The philosopher's teaching started in a monastic order. From there it slowly spread to a greater following and eventually changed the culture of the time.

What was developed in closed Pythagorean circles eventually became a general capacity. Nowadays the science once restricted to the Greek elite is at everyone's fingertips. The conceptual understanding of the Pythagorean theorem, once reserved for a highly trained elite, is now mastered by school children.

The same applies to Buddha's impulse in the East. Before his time, conscious compassion was virtually unknown. Love was linked to blood ties and compassion was practised within the circumference of the familiar. Beyond these parameters the ancient world was laced with a matter-of-fact cruelty foreign to our sensitivities. Compassion developed only gradually, and Buddha was the first to effectively embody this capacity and conceptually formulate its universal appeal. As with Pythagoras, compassion passed from the exalted saint to immediate followers and from there to an ever-widening circle of laymen. As Buddhism spread, compassion followed in its wake. It grew through direct transmission as well as through the subtle means by which capacities developing in one part of the earth appear in another: the proverbial 'hundredth monkey effect' on a global scale. The relatedness of these impulses can be seen in the simultaneous emergence of conscience as a weaker form of compassion in the West, and of a first, but intermittent, flowering of science in the East. By and large compassion remained foreign to Occidental affairs, science peripheral to Oriental concerns.

In the West the birth of conscience immediately follows that of science, and is well documented in the myth of Orestes and its dramatic treatment by the Greek poets Aeschylus and Euripides.[1] Conscience becomes the Western substitute for Eastern compassion: a fainter, less immediate version of its Eastern relative. It is a kind of compassion after the fact. Pangs often strike after deplorable deeds. Even if conscience stirs before, it does not prevent us from being tempted in the first place. Reminding us of our moral relatedness, conscience implicitly admits to our separateness. Compassion, by comparison, is a spontaneous, immediate, unpremeditated act. It is a manifestation of an already existing unity that we need not be reminded of.

In the West, conscience remained a thin layer of morality behind a thick skin of insensitivity: ancient Greeks had no more qualms about the sacking of Troy than did the Romans about the destruction of Carthage. These were matters of ongoing pride rather than belated remorse. Slavery was an undisputed fact in antiquity. Spartans and Athenians built their wealth on the exploitation of slaves. Nor were the Portuguese, Dutch, Spanish and English much concerned about their atrocious colonial exploits. New World pioneers dislocated native Americans just as readily as white Australians uprooted Aboriginal culture. But that is changing. A new chord has been struck on the heartstrings of this world. Conscience is coming of age in the West (and science is taking a firm hold in the East).

When conscience matures it approaches compassion. This new compassion now grows not just in individuals and in groups, but as a global faculty too. As it spreads over time and space, the earth is becoming an area of communal care. Our concerns about the third world, about hunger and poverty and social injustice, our care for disappearing forests and changing climate, for global warming, rising sea levels and spreading deserts are indicative of this change. Former heroes of conquest are overshadowed by heroes of care. Restorative justice, appreciative inquiry, non-violent communication and many other compassionate means are becoming widely applied social technologies.

I see this globally emerging compassion as the capacity of the earth to regulate its moral sphere as it does its biosphere. To aid this we must become not only sensitive to the suffering of the world, but cognisant of its causes. The participatory science of Goethe and Steiner does both: it co-suffers by overcoming separation and cognises through active imagination. Historically, science and compassion developed independently to come into their full strength. Now they need to unite and inform one another. The rapid increase of NGOs, community groups, charitable foundations and environmental organisations all over the world testifies to the transformation of conscience into compassionate acts. At the time of Pythagoras and Buddha the world was only loosely linked. East and West went along separate paths. Now everything is global. West and East have merged. Eastern spirituality has made its way to the West. Modern Western technology has changed the East. This meeting must not remain superficial.

Acknowledging the parallel between Western science and Eastern mysticism is not enough. What matters now is to apply inner capabilities to outer phenomena. This is both an historic and an environmental necessity. Science cannot continue without a moral dimension. Compassion is called for in the way we investigate nature.

15

GLOBAL EMPATHY

Long before I became aware of colony collapse, this book announced itself through my series of bee-related artworks, among them *Buddha in Honey* (see Figure 11, overleaf), the visual that inspired the notion of compassionate ecology. Much contained in this volume is an extension of this artwork, an attempt to express in words what the picture contains: a fusion of compassion with environmental care through the bringing together of the image of Buddha with the substance of honey.

I am not a Buddhist, and I intuit artworks rather than premeditate them. In hindsight, however, the combination of Buddha and honey seems appropriate for a number of reasons. Firstly, the tenets of Buddhism promote exemplary care for the environment. The following passage from Heinrich Harrer's *Seven Years in Tibet* testifies to the compassionate attitude of Tibetan Buddhism:

One cannot close the heart to the religious fervour that radiates from everyone. After a short time in the country it was no longer possible for one thoughtlessly to kill a fly, and I have never in the

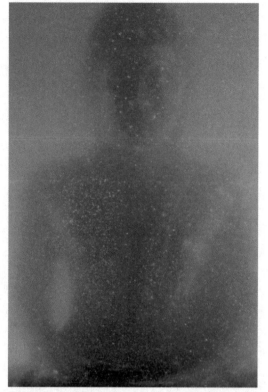

Figure 11. *Buddha in Honey 2*, Horst Kornberger, 2001.

presence of a Tibetan squashed an insect which bothered me.
The attitude of the people in these matters is really touching.
If at a picnic an ant crawls up one's cloth, it is gently picked up
and set down. It is a catastrophe if a fly falls into a tea cup...
Typical of this attitude towards all living creatures was a
prescript issued in all parts of the country to persons engaged in
building operations... It was pointed out that worms and insects
might easily be killed during the work of building, and general
construction of buildings was forbidden. Later on when I was in
charge of earthworks, I saw with my own eyes how the coolies
used to go through each spadeful of earth and take out anything
living.[1]

I am always moved by this passage. Yet I am also aware that such care is not transferable to the rest of the world. Flies will be swatted and pigs turned into pork. The singular fly, bee, cow is beyond the compass of Western compassion. But the bee in all bees and the cow in all cows must not remain outside. The essentials must enter the radar of global empathy. Unless we, personally and collectively, feel responsible for the well-being of every species, abuse will continue.

The world needs a good dose of Buddhism. We cannot import Buddhism as it was, but we can resurrect its essence in new and appropriate ways for our times, and independently of religious preference. We can do this when, as a civilisation, we practise non-violence towards every species. In other words: while cows will still be eaten, the treatment of cows will improve. The essential cow will be recognised for her needs and treated accordingly. Compassion toward the essential cow, bee or cod is the Buddhism required by the world today.

Secondly, Buddha is a globally accepted metaphor for meditative inquiry. Submerged in honey, the meditating Buddha points to the inner awareness we ought to bring to our outer phenomena. This is the meditative-compassionate element, pioneered by Goethe and those who followed in his wake, with which we need to infuse our sciences.

Thirdly, Buddha and honey are both related to essentials. Buddha is not so much a name as a designation for a human being who has left all incidentals for the sake of the essential. In honey, nature itself has become essential. As a substance it surpasses milk, oil, wine, salt and incense in its sacred appeal. Of all substances it is perhaps the most essential and therefore the most suited to represent compassionate care for the essential dimension in plants, beasts, groves, wells, rivers, landscapes and ultimately the earth itself.

These second and third reasons complement each other. The meditative element brought to externals balances the objective approach to inner phenomena, brings imagination to science and science to imagination. The latter is just as important as the former, as our care for essentials must not stop with tangibles. It must include the endangered species of the inner life, most particularly the imagination. The latter is to the modern mind what a key species is to the environment.

Millennia ago, the imagination was the natural habitat of the pre-individual mind. Our intellectual age has pushed it to the brink of extinction, where it barely survives in isolated pockets of art.

The environmental Buddhism of our time is inevitably linked with the development of the imagination. Buddhism pioneered what was in its time a highly conceptual doctrine amid a population still living in an imaginal mindset. In the intellectual climate of today, environmental Buddhism must reverse this trend and spearhead the imagination as a cognitive tool. Unless the imagination is restored there will be no lasting solutions for either bees or environment.

Luckily the imagination has not become extinct. Nor has it just survived in artists and cultural creatives. Everyone has a fair share in this capacity, though few put it to use. Most of us pay too little attention to the imagination's fine print on the margins of the mind. Few realise that our thoughts are slightly interwoven with and, on closer observation, even preceded by imagery. To work with the imagination preceding thought requires the kind of training proposed by Goethe and Steiner. To work with the images that accompany the everyday mind is within everyone's reach.

The fact that we all can understand metaphor testifies to this ability. This understanding is the beginning of imaginative literacy, the globally emerging capacity to understand by means of picture rather than concept. This ability underscores the first step outside the prevailing paradigm and the hold it has on our mind. It is a capacity we already have but little use. Once encouraged however, it quickly comes to life: it is like learning a language that we already know. Once learned, the world begins to speak in pictures. We begin to understand the fuller implication of our bee crisis. Unless we begin to take metaphors as seriously as storm warnings by the weather bureau, none of these problems will be solved. We will not find solutions by applying concepts alone.

Simply stopping queen bee abuse is not enough. Nor will reverting to preindustrial beekeeping reverse the harm inflicted. To think that the bee crisis can in any way be remedied within existing parameters is illusory. We need to think outside the bee box. And for this we need to understand metaphors.

16

BEEHIVE METAPHORS

For so work the honey-bees,
Creatures that by a rule in nature teach
The act of order to a peopled kingdom.
They have a king and officers of sorts;
Where some, like magistrates, correct at home,
Others, like merchants, venture trade abroad,
Others, like soldiers, armed in their stings,
Make boot upon the summer's velvet buds,
Which pillage they with merry march bring home
To the tent-royal of their emperor;
Who, busied in his majesty, surveys
The singing masons building roofs of gold...

William Shakespeare, *Henry V*

Metaphors are important because they engage more than the intellect. They are the language of relation. By expressing one thing by means of another they assert the communality that links everything. A true metaphor, however, is anything but arbitrary: if I call a pen a lightning rod for new ideas, I have not succumbed to fancy. I have simply revealed a meaning the pen already encompasses. Real metaphors characterise, enlarge and reveal more of an idea. They bring to consciousness what we were already vaguely aware of. Metaphors clarify and yet retain mobility by allowing multiple meanings. In other words, they are alive, like nature.

Scientific concepts too are metaphors, albeit restricted, exclusive, somewhat shrivelled ones. They apply to one phenomenon and not another. The formula $a^2 + b^2 = c^2$ applies to right-angled triangles and not to oblique ones. Such tight metaphors suit physics, inorganic chemistry and mechanics. They are perfectly appropriate for the making of machines but are desperately inadequate to life. Their precision necessitates an isolation that disaffects the world. Poetic metaphors, on the other hand, speak the primal, visceral language of relationship because they assert connectedness. They are conceptual socialites, providing the connective tissue that weaves isolated phenomena into a meaningful web, connecting more of us with more of the world.

The interrelating metaphor was the indispensable tool of the mythopoetic ecologies that existed in the past. These ancient forms of environmental care were inseparable from sacred tradition and the metaphoric language of myth.

The oldest of these ecologies exist in Australia. Here, site-specific creation myths, called songlines, map the land and are maintained by those initiated into lore and law. These mythic maps describe the deeds of ancestral divinities who created the land and all it contains. Thus waterhole and rock, cliff face and creek, wombat and bee are linked by a common body of myth that assures familiarity and care.[1] Here, each tribal group relates to a creative ancestor and a corresponding tract of land. This makes each group responsible for a specific part of the land and the totemic animal associated with it. If the wombat group lacks honey, they ask the bee group for an increase of bees. The bee group then approaches the ancestral creative entity of the hive (the bee in all bees) through appropriate rituals. The need for honey turns into a social act and the social act into spiritual practice. The result is a culture of care and sustainable ecology.

In this system one group feeds the next. The custom resembles that of bees, who do not eat in solitude but feed each other inside the hive. There are other parallels too: group organisation corresponds to bee organisation in terms of task groups, and the creative ancestor regulates land fertility in the same way the queen bee regulates fertility of the hive. Beehives have been taken by many cultures as a symbol for their ideals. In the traditional society of Australian aborigines, the ideal was reality.

In later cultures, bees become associated with the sacred and divine (to Aborigines everything is sacred) and often to the creator gods or goddesses themselves. In ancient India, for instance, the blue bee resting on the lotus of life is the symbol of the all-creator Vishnu, father of all the gods and preserver of the world. The bee emblem thus appears on the forehead of Krishna, Vishnu's most loved avatar.

Egyptians too linked the origins of bees with the creator god himself: 'When Ra weeps again, the waters that flow from the ground turn into working bees.'[2] (Ra was to the Egyptians of early patriarchal times what the creative mother goddess was to earlier societies and what nature is to us today.) Thus the bee hieroglyph served as a symbol of the pharaoh, Ra's earthly representative and ruler over the land of plenty.

In ancient Greece, the relation between bees and the creative mother goddess (the representative of nature) was a matter of fact. Priestesses of earth goddesses such as Artemis, Rhea, Cybele, Demeter and Persephone were commonly called 'Melissae', or bees. Not surprisingly, many female deities representing nature were depicted as bees. On the plate on Figure 12 (overleaf) the great mother goddess is half woman, half bee. Among the many deities representing nature, Artemis (see Figure 13, p. 93) was perhaps the most prominent. She was to antiquity what Gaia is to our time. The statue reveals the figure of the Ephesian goddess covered with animals: she appears as a life-giving queen bee amid the world-hive of creatures. Bees feature prominently and in some depictions they appear all over her body, representing the same activity on the sculpture as they do in nature.

Even in times of growing intellectuality bees retained their sacrosanct status. Scientifically minded Aristotle still assigned divinity to bees:

Concerning the generation of animals akin to them, as hornets and wasps, the facts in all cases are similar to a certain extent, but are devoid of the extraordinary features which characterise bees; this we should expect, for they have nothing divine about them as the bees have.[3]

Among commercially minded Romans the divinity of bees began to fade in favour of the hive as a symbol for the perfect society. Over the centuries, beehives were appropriated by various groups. Roman senators, church dignitaries and American Puritans saw beehives as a fitting illustration of their own ideals. To the Roman republican they represented a state upheld by communal effort, to the clergyman preordained hierarchy, to the Puritan diligent work. Shakespeare, living in Elizabethan times, describes them in terms of monarchy in his *Henry V*.

But all these metaphors are out of date. Severed from the initial experience that stood at their inception, they become clichés. And many have turned into cover-ups. Today, the proverbial 'busy bees' mask industrially enslaved hives, and 'sweet honey' the sour practices applied in its production. The real metaphors today are written in the prose of profit and loss and printed in the definitions of science. Cast into Latin, bees have become *Apis mellifera*, a species classified to abstraction. Scientific jargon is the language of separation, of objectified indifference.

Figure 12. *Gold plaques embossed with winged bee goddesses from Camiros,* Rhodes, seventh century BCE.

Figure 13. *Diana (Artemis) of Ephesus statue* in the Candelabra Gallery, Vatican Museums.

And indifference invites abuse. Scientific conceptions combined with economic imperatives have made the once sacred bee an insect exploited for the sake of honey and pollination. To address this state of alienation we need a new language, so that meaningful, emotion-filled pictures come to mind whenever we see a bee, taste honey, touch beeswax, come across a thriving hive. Such pictures are important for the communal psyche, because they disallow abuse. Unless we use our imagination to recover the sacredness of the bee, the numinous dimension of honey, the symbolism of the hive, there is little hope for bees.

Here bees offer us the very metaphors we need today: they are the master ecologist of nature. Bees serve the whole of nature while serving themselves. Their work is universally beneficial and finely attuned to the environment. Hive and sustainability must become synonymous. Bees and ecology should be linked in our mind as they are in nature: the welfare of bees is a global barometer of environmental care.

A second important metaphor is that of master economists. Careful division of labour and complex organisation allows them accrue honey beyond their own needs. Every beehive offers a picture of a successful economy. And most important of all they are able to do both: their economy does not impact on their ecology; their ecology does not impede their economy. On the contrary, the more economically successful they become, the more of an ecological benefit they are: a mutually enhancing ecology and economy is the new global metaphor of the hive.

But we must not stop with the positives. These alone cannot help the crisis. The most crucial bee metaphor for our time is colony collapse. It is the metaphor that nature herself provides. Unless we understand this metaphor in detail we cannot hope to remedy its causes. We have looked at these causes from a conceptual point of view in the beginning of this book. Now we must look at them once more by means of imagination. Only by understanding the metaphoric significance of bee frames, wax foundations, artificial breeding and forced insemination will the full dimension of the current crisis reveal itself.

17

BEE FRAMES AND MIND FRAMES

The question is not what you look at – but how you look and whether you see.

Henry David Thoreau, *Journal*

Nature is the symbol of spirit.

Ralph Waldo Emerson, *Nature*

Rectangular bee frames and prefabricated wax foundations mark the beginning of modern beekeeping. Both result from the combination of scientific insight and commercial rationale, and introduce a foreign element into the hive. Bee frames came first and forced the naturally heart-shaped comb into rectangular formations. In nature, form and function coincide. The nests of birds and the warrens of rabbits fit the lives of their occupants. The same is even more true of bees, who literally exude their buildings from their body. The shape of the comb expresses their nature. It is a shared artwork that begins in wax production and ends in sculptures that balance the properties of honey and wax. To squeeze this artistic, organically rounded structure into rectangular frames impedes the self-expression of bees.

Suppression becomes oppression when prefabricated wax foundations are forced upon bees. To take away a good part of the bees' opportunity to make wax oversteps the mark. The imposition of frames affects the shape of what is built; a prefabricated foundation affects the process of wax production itself. Bees are made to exude wax. It is neither work nor idle occupation, but part of their nature and their environmental activity.

The source of beeswax is the sweet liquid that blossoms exude, produced in the living refinery of the plant. This nectar is separated into honey and wax in the bee's body. That which becomes honey goes into the stomach, to be later regurgitated. That which becomes wax enters the bloodstream and is eventually exuded in wax scales on the abdomen of the bee. These scales are removed by fellow workers, who chew them. When the wax is finally bonded to the comb it has passed through the body of one bee and the mouth of another.

The process of wax production is intriguing. What is first separated in the bee reunites in the comb: nectar turned honey is stored in nectar turned wax. Comb, however, is more than a storage facility. It is the communal body of the hive: an internal shell, an artwork, a home, a store, a stomach and a womb.

To fashion honeycomb is the ancient prerogative of the hive. But if you look at it with profit in mind, comb-building is counterproductive. Time and energy lavishly spent on architectural construction could be used elsewhere. Beekeepers therefore introduce artificial wax foundations to speed up the construction of cells. Bees are thus 'freed' from the tedium of waxwork, and the time spared on wax-making is spent on honey production instead.

Here, commercial logic and eco-logic clash. To the hive, the capacity to build comb is synonymous with the capacity to build its own body. Inserting pre-made foundations stymies the creativity of the individual bee and suppresses the bonding into a communal body of wax – yet another attack on bees' intense relatedness.

At this point, the plight of the honeybee can supply us with metaphors for our own predicaments. What we do to bees we do to ourselves when we adjust the all-rounded universality of the human mind to the rectangular frame of the intellect and allow the fixed foundation of paradigms to limit our collective self-expression.

Just as it is bees' nature to exude wax, it is human nature to exude new ideas, develop innovative approaches and find fresh solutions. Bees build their comb from their bodily wax; we build our civilisation from the malleable wax of our mind. What individuals think, imagine, invent and produce becomes the communal body we call civilisation. And this civilisation is as stuck in the regulating rationales as the comb inside rectangular frames, and as hemmed in by paradigms as bees are by pre-made cells. Within our stifling mind frames, collective potentials stay unrealised and individual capacities untapped. All paradigms change over time. They are temporary perspectives and therefore necessarily one-sided (the 'collective representations' that Owen Barfield has so brilliantly explored[1]). In focusing on one aspect of reality, they ignore another. Paradigms cannot but imprint this in-built partiality on all they produce and their most iconic achievements are therefore the most one-sided: the milestones of one age become the millstones around the neck of the next. Newton's gravity-powered world-machine is among the heavyweights. This theory perfectly explains why the apple falls to the ground, but not how it got up there in the first place.

Darwin's theory, in which evolution is propelled by competition and chance, is another heavyweight. Evolution, of course, is a fact (and as inspiring to the imagination as to the intellect). The theory behind evolution, though, is one-sided fiction: it illuminates the role of competition, but obscures that of collaboration. The latter plays an equally, if not more important role in the totality of nature. Both Newton's and Darwin's theories are powerfully convincing not because they adequately describe the workings of nature, but because they resonate with the nature of the mindset that produces them. This makes their influence pervasive and exceed the sphere of science.

Newton's mechanical line of thought revved the motor of social theory and thus shaped reality. Darwin's theory did the same to an even greater extent. There is, however, one major difference between the contributions of Newton and Darwin. The weight-powered universe remained a conception. Darwinian theory became a story. And stories are powerful.

In my book *The Power of Stories*[2] I explored the impact of stories on civilisations. I found that myths were singularly influential on the people who possessed them: all cultures fulfilled the content of their myth in their later destiny.[3] Creation myths (stories that give an account of the becoming of the world) are particularly potent in this respect. Illuminating the past, they shape the future, setting the tone for all that is to come. Today, Darwin's theory is the universally accepted creation myth, setting the tone of our current civilisation. It has no less influence on our global civilisation than the old myths had on their cultures.

In this scientific creation myth the world begins with an unlikely accident. It evolves by random combination and is powered by universal competition in the struggle to survive: the moral inside this depressing tale is the universal warfare of everything against everything else.

If ancient myths have become reality, Darwinian myth will do the same. Indeed, it cannot do otherwise: it has long since ceased to be a theory and has become unquestioned reality. From the moment we accept its premises it not only excuses competitive behaviour but also encourages it. The theory forces the conclusion that unless we too become competitive ourselves, we will not survive. Simply by believing this conception, one makes it come about. The myth creates what it proposes and thus proves its initial proposition.

Looked at more closely, it is a blow to human compassion, similar to that in the school assignment described in Chapter 10. This most depressing narrative pervades culture, influences politics and shapes social reality. Darwinism is foundational to contemporary thinking, instrumental to science, influential in psychology and ubiquitous in economy, politics and education. It is most powerful where it is least observed: in the subcutaneous layers of the mind, in the preforming of opinion, in the subtleties of human interaction. Its presence is blatant in popular media and violent computer games.

Darwinism is to contemporary culture what preprinted wax-foundations are to commercialised bees. And as with bees, the paradigm behind Darwinism impacts on our collective self-expression when we fit our social realities to Darwinian specifications and plan our collective future on competitive theories rather than compassionate practices. What is detrimental to humans inevitably becomes destructive to nature. Human beings can articulate their concerns, protest against pain, act against abuse, but nature has no other tongue than her sufferings. She speaks through the loss of soil, the disappearance of forests, the depletion of fisheries. She cries out through failing climate and missing bees.

Through colony collapse, nature illustrates the ineptness of competitive paradigms. The beehive is nature's showcase of collaboration: bees support each other and everything else. It is inevitable that thoroughly collaborative bees are the first victims of a competitive paradigm. It can hardly be otherwise. Reductionist science creates the know-how of abuse, and competition the economic realities that force us to apply it to nature. Buying into this paradigm, we declare war on every species. We must subjugate, exploit and eventually destroy for our own survival. In nature, competition is always part of a greater collaboration. By contrast, our one-sided belief in competition stands behind stressed cows, propped-up pigs, abandoned beehives, failing forests, missing topsoil and an everywhere ailing earth.

Frames and foundations alone, of course, do not force bees to extinction, nor do our paradigms by themselves undo society. Mindsets cannot be avoided and are, to a certain degree, even necessary (as perhaps bee frames are too). We cannot help having paradigms. But we can help paradigms having us – particularly when they are no longer helpful. Then it is time to swarm our mindsets.

The Beehive

For days the bees clot
on the outside of the hive
ripening for the swarm

the moment comes as a
thickening of the light,
they pitch above in a wild
and precise choreography
a laval flow of bees
spills from the opening
and is sucked into
the spiralling thrum

they expand into a planet
of droning atoms,
ascend into suburban space

as they pass over the house
their wordless mantra
intones the interval
between a warning
and a blessing

it enters the dark chord
of the body,
weighs the heart against
the pure scale of honey
and passes on.

Jennifer Kornberger

18

THE CHOREOGRAPHY OF CARE

I still remember the first time I saw bees swarming. It was in spring, around noon, when the sun was close to its zenith. There was unusual commotion in front of the hive. Invisible floodgates opened and bees poured forth. More and more hurled themselves into the open, drawn irresistibly into the expanding whirlpool. For minutes I stood mesmerised, the air thick with bees.

Eventually the frenzy calmed, the airy hive thinned and folded into a cluster on a nearby branch. The queen had found a place and drew her followers to her. Bees grew around her like a living, shivering fruit. The small bundle soon swelled to formidable size, suspended between the future and the past. Soon all the bees had gathered; only a few scouts sallied forth in search of a hollow tree, a crevice between walls, an entry into the dim interior of a roof.

The hive that had so vigorously expanded just a few moments ago now contracted into a state of rest. I knew that in one or two days the colony would find a new abode and commence its work: workers would surround their matriarch with elaborate rituals of attention, gather honey, build their palace of wax and rear brood. Back in the old hive, it was bee business as usual. A new queen was preparing to rule the remainder of the hive. She would soon leave for her short nuptial flight and return with seed enough to last her for years.

When a hive swarms, the old queen leaves and a portion of her people follows her in a dedicated trance. The one hive becomes two. In a year, sometimes less, the hive will swarm again. Swarming is part of the reproductive cycle of the hive, a kind of cell-division on a large scale. The result is more hives and more bees.

Traditional beekeepers watched their hives carefully for signs of impending departure, ready to catch the new swarm. Drunk with honey to last the lean days of transition, the bees are dazed and docile. The whole cluster can easily be shaken into a box and the box later emptied into a proper hive. Thus the colony is kept and the apiary increased. A hive can swarm more than once. In one season a colony might swarm two, three or even four times. On rare occasions there may be even more swarms, or casts. Beekeepers then speak of swarming fever. Swammerdam observed a hive swarming thirty times after a fierce winter. But such exuberance is rare.

Swarming is a courageous act, a wholehearted embrace of change and rejuvenation. Every year, every hive follows this ritual and dares the unknown, risking deprivation, discomfort, danger and sometimes death. The old queen departs, leaving her palatial abode, her stores of honey and pollen and brood for a new (as yet unhatched) queen and her people.

Queen bees do not take easily to rivals. There is always one and only one queen in a hive. The old queen leaves before the new queen arrives. Once hatched, the new queen immediately dispatches all of her would-be rivals. With the consent of her subjects she will gnaw into the capped brood chambers of her royal relatives and sting them to death. If, however, the hive has 'decided' to swarm a second time, the new queen will be prevented from venting her jealousy and must endure the maturing of other queens. But before her rival is born, she, like her predecessor, will set forth with a portion of the hive.

The order of royal succession has long puzzled science. The hive somehow 'decides' how often it wishes to divide, and protects or sacrifices its royals accordingly. There seems no rational pattern for these decisions.

I doubt that the decision to swarm is dictated by the hive alone. I believe it is brought on by the vagaries of the environment. Swammerdam associated the excessive thirty swarmings of his hive with an extremely cold winter that had killed many colonies. The surviving hive must have been extraordinarily vigorous. For its own survival, such profuse reproduction might not have been necessary: very likely the hive attempted thirty casts for the benefit of all bees in the area. I believe swarming is generated by the bee in all bees that regulates the number of colonies in a given area through the succession of its queens. A continuum that works from the environment into the single hive and from there into the reproductive habits of queens makes immediate ecological sense. Steiner points to the existence of such laws in his bee lectures:

In nature there are remarkable connections between everything. And the most important laws are those that people can't comprehend with their ordinary understanding. Of course, these laws have a little bit of flexibility. Take for instance the balance between the sexes on earth. There is never a completely equal number of men and women on earth, but the number is approximately equal. Over the entire earth the number is approximately equal. The wisdom of nature controls this. If the time should come… that humanity will be able to arbitrarily determine the sex of children, then the whole balance will come into disorder. The situation is such that when the population in a certain area has been decimated by wars, the population becomes more fruitful and more children are born. In nature every deficiency, every shortage calls forth for its opposite.[1]

More males were born after World Wars I and II, a fact that remains unexplained by conventional science.[2] To the imaginative mind such laws are as obvious as they are necessary. Nature seeks balance on the scale of the whole. When we recognise that bee reproduction is a localised balancing act, based on communication between environment and hives, we can see that the bee in all bees provides an area with sufficient hives and diversity of bees. Through bee reproduction the environment regulates itself. Pollination stimulates plant propagation, and plant propagation affects everything else: soil, insects, birds, and mammals. To prevent swarming is to sever connectivity between bees and environment. Whenever we interfere with bees, we interfere with the whole of nature and harm them both.

To the intellect, bees are an addition to the environment, hives are separate entities and queens interchangeable birthing machines. To the imagination, however, environment and bee form an inseparable whole. Swarming is part of the blood circulation in the body of nature. Its prevention is to the environment an intellectual heart attack.

19

SWARMING PARADIGMS

Imagination is more important than knowledge. For knowledge is limited, whereas imagination embraces the entire world, stimulating progress, giving birth to evolution. It is, strictly speaking, a real factor in scientific research... We cannot solve problems with the same mindset that created them.

Albert Einstein

Reproductive swarming is to the hive what the production of new mindsets is to human culture: a daring act of mental rejuvenation regularly required by the environment. If this mental swarming is suppressed, the vital link between our collective potential and the needs of the environment is broken. Through colony collapse, our environment is calling us to swarm our paradigms and restore the biodiversity of the global mind as a prerequisite to restoring the biodiversity of nature. Such large-scale transformations require new capacities.

The development of these capacities is the task of education. Because swarming is to the hive what learning is to society, bees can teach a valuable lesson.

Some bees, bumblebees for example, don't swarm at all. This roughly corresponds to a level of human learning that requires little change. We learn more of the same and little that challenges what we know. The knowledge we already have remains intact. No real mental swarming occurs.

The social bees of Australia (whose polar relationship to the European honeybee has been explored by Brian Keats in an interesting essay[1]) swarm in the way we might expect: the young queen leaves the hive and the old queen remains. This responds to a level of learning where the arrival of new insights changes old convictions. This occurs in scientific breakthroughs and major advances in art. The discoveries of Swammerdam and Dzierzon, for instance, radically altered the way bees were thought of, but not the way of thinking. The prevailing paradigm remained essentially unaltered. (The new queen leaves without altering the hive.)

The European honeybee, *Apis mellifera* (the master ecologist among insects) swarms differently. Here, the old queen leaves and the new queen stays. This ensures generous support for the new queen and allows space for a different style of governance (the new paradigm inside the hive). It also means that the old queen and her followers continually expose themselves to rejuvenation and change. The new population is encouraged and the old kept flexible. What otherwise would become static remains capable of serving the environment.

If we were to translate this in terms of human learning it would amount to a dynamic collaboration between tradition and innovation: fresh and challenging ideas and paradigm-changing attitudes (the new queen) would be given ample space to unfold their full potential. And the already existing mindset (the old queen) would welcome them as necessary rejuvenation.

But this is not how it is today. Education strongly favours the first (the bumblebee model), and research the second style of learning. The third is not only discouraged, but systematically suppressed by the way we educate. We do this by superimposing the established mindset on ever earlier ages. This forces adult intellectuality on new and nascent potentials and so prevents them from developing. To change this, outdated modes of learning, appropriate in the early, assertive phase of the scientific mindset, must be left behind. To learn as much as possible in the shortest amount of time at an ever earlier age must be seen for what it is: the militant means through which the current paradigm reasserts itself to the detriment of new developments, crushing the forces of renewal where they are most likely to occur: in children.

Childhood in its real, essential, unadulterated form is continuous evolution, revolution and change, a period of intense learning, quickly changing phases and profound shifts. Every year, children encompass new mental and emotional horizons and explore different ways of being. New attitudes are lived and worked through. What is accomplished in one year becomes foundational for the next. All this makes children natural geniuses and change-makers, ambassadors of the future who bring the capacities the world needs.[2]

To help them preserve these capacities into adult life, children must be allowed to experience childhood to the full. Intellectuality must be held at bay and each phase of learning cherished for the gift it brings. The solution here is a non-rushed, age-appropriate development of all human potential rather than taking the fast track to adult intellectuality. Schooling must be less about acquiring specific knowledge (which is quickly outdated) and more about developing abilities that promote life-long learning, such as to adjust, change, create, innovate, improvise, empathise and imagine (a principle Steiner applied in Waldorf education).

This in turn requires educators who can model these capacities, who are free to adjust, change, create, innovate, improvise, empathise and constantly reimagine and reassess what they do. In other words, it requires teachers who are able to respond to the educational situation at hand rather than to outcomes prescribed by authorities. Such teaching is stifled by hierarchical models that make teachers administrative clerks. New capacities need a bottom-up approach prompted by the true needs of children rather than by political agendas or economic imperatives.

All this requires a thorough reimagination of educational models. Here, bees are our double mentors: they teach us through their demise that radical paradigm change is necessary and through their swarming how to bring it about.

20

COMPASSION COLLAPSE DISORDER

Compassion is the basis of all morality.

Arthur Schopenhauer, *On the Basis of Morality*

Education, of course, is not confined to schools. We learn throughout life, and most intensely in early childhood. All children are geniuses of learning, as their quick acquisition of language testifies (a theme brilliantly explored by Hungarian logosopher Georg Kühlewind[1]). Their genius is best expressed in the mobility of the imagination: children easily turn a block of wood into a house or car or cat or cow. They are not yet paradigm-locked and still have what adults have lost: the ability to think in flexible pictures. Studies in divergent thinking have shown that they can easily imagine two hundred uses for a paper clip, while adults struggle with ten.[2]

This imaginative 'wildlife' of childhood is the playground for the very capacities we need in later life: here, imagination secures our choices by allowing us to find five solutions to a problem rather than one. Without imagination there is no creativity, and without creativity no change. We remain locked in the box in which we are bred. This applies particularly to mindsets, which need a collective surplus of imagination in order to effectively shift.

To develop compassionate ecology, the imagination is essential: mature imagination links accurately where childhood fancy relates arbitrarily. This becomes the capacity that Goethe systematically developed into a cognitive tool and that Steiner elaborated upon: the ability to co-imagine with nature rather than conceptualise about her.

Naturally, the ability most needed to change our paradigm is most fiercely persecuted by it. The imagination is marginalised in education, minimised in culture and constantly attacked in the years before school. Yet it is in these years that much learning occurs: ecology is taught by nature, community by the family, love by the parents and imagination by play. This foundational learning is endangered as technology increasingly substitutes for nature. Today, media entertainment replaces family life and computer games substitute for creative play. In consequence the imagination suffers severely: in screen mode, children consume the pictures they ought to produce. Just like bees who forget the art of comb-building when foundations are forced upon them, children forego the production of inner pictures when they are superimposed from outside. This will limit their choices in the future, impede their ways to think differently and stymie the vision they have for themselves and the world. Above all, paucity of imagination will impact on empathy, which depends on the ability to vividly picture and thus feel the plight of the world. Undermining this ability, media exposure indirectly assaults compassion itself.

Today, this indirect assault is rapidly becoming a direct attack. Compassion is about real relationships, and these have their roots in the child's bond with their mother or other committed caregiver. The child lives within the mother's emotional field, partakes in her moods, sentiments, thoughts and in everything she experiences. This first relationship is the matrix for all others. In other words, children partake in their mother as we ought to partake in the world. The feeling of oneness with the mother later turns into the feeling of being related to the world: that is to say, the capacity of compassion.

In the last two decades technology has begun to interfere with the mother–child relationship. Mothers are increasingly replaced by screens, electronic pets and robotic babysitters. These electronic substitutes are intelligent without warmth, clever without compassion, brilliant without concern. They compute without understanding, move without feeling,

speak without meaning. They are present without being th
a relationship, they prevent it from developing. The link to be

Exchanging authentic for imported queens is to the   ..iaC
supplanting mothers with machines is to a human being: the former
is an attack on what is most essential to bees, the latter an assault on
what is most essentially human – our ability to relate, care about and
empathise with family, friends, fellow human beings, the nature we live
in and the world we inhabit.

What causes colony collapse will in time cause the breakdown of
civilisation and much else. When compassion is at stake, humanity is
in danger and the earth at risk. Genetic manipulation will take this yet
further. The production of perfect bodies and super-bees is almost a
given. Like train tracks merging in the distance, the parallels between
bees and humanity meet on the nearby horizon of genetic engineering.
At this point, metaphors disappear. What happens to bees and what
happens to human beings will be the same.

Media invasion and mothering robots both affect capacities. Genetic
engineering goes even further. It affects the very instrument on which
these depend: the human body, the structure of the brain, the shape and
functions of organs. We will live with generic bodies and with brains
made by the same mindset that drives bees to extinction.

The 'engineering' in genetics clearly points to the model of the machine.
Mechanical thinking incarnates itself in the body and the paradigm becomes
flesh and bone, with human beings instrumental in agendas that endanger
the world.[3] We have looked at mind frames and the ways in which they limit
culture. We have seen how they reinforce themselves through education and
thus constrain the possibility of change. By means of technology, the current
paradigm extends its influence to the formative years of early childhood.
Usurping the imagination and obstructing the development of compassion,
it attacks the very capacities that could bring about transformation. Even the
possibility of altering the human body is within reach.

This is the direction in which we are heading. Unless we change it.
And to change it, we need imagination.

Extinction of the Phoenix

It happened fast.
One was trapped in fact.
Another talked over and done with.
A third one appointed to the past.
Others were shot
by the irreversible arrow of time.

The last survivors died
under the sceptoscope of doubt.
One escaped
into the safety of imprisonment:
a legendary pet
inside the cage of myth,
a fire-extinct species
in an imaginary Zoo.

He has no chance
and yet he flaps his wings
as if he were to fly.
There are no flames
and yet he still looks burnt.
There's no escape
and yet he pecks the locks
as if a door would open.

His upkeep, though, is easy.
He feeds on words
and other crumbs of resurrection
and drinks the tears he sheds.
He weeps, it seems,
for all his captors' sake
who know not
what's inside
and what's outside
of a cage.

Horst Kornberger

21

IMAGINAL LITERACY

The paradigm we have – or rather the paradigm that has us – fiercely governs the way in which we think. Its rule is supreme, but not total. What is suppressed in upper layers of the mind surfaces in others; what the intellect rejects the imagination is ready to reveal. In a thoroughly intellectual world where censorship happens in everyone's mind, this is not easy. The only place, it seems, is 'undercover', in children's books and fantasy novels. Here alone the imagination has free range.

Much of what is said here has appeared in works such as C.S. Lewis's *Chronicles of Narnia*, Michael Ende's *The Neverending Story*, Philip Pullman's *His Dark Materials* trilogy and of course, J.K. Rowling's *Harry Potter*. Their popularity is proof of their efficacy. Rowling's series put a mirror in front of the mind and painted a portrait of the paradigm war behind scenes: Harry Potter is the likeable boy next door. He is no superhero, but amiably human. He has, however, an unusual gift. He is the magical–imaginative child within intellectual reality. Hence his ultra-conventional step-parents, the Dursleys, suppress his capacities in every possible way.

Harry Potter exemplifies the destiny of the imaginative child in all of us. And like all of us, he has to wrestle with a paradigm that seeks total control. For Harry, this is Lord Voldemort. The evil magician knows that Harry (the imaginative child) will challenge his own bid for absolute power. Thus Voldemort, just like the current paradigm and its attack on imagination and compassion in young children, attempts to kill Harry as soon as he is born.

Harry survives through the sacrifice of his mother to face a continual struggle with his step-parents on this side of conventional reality and a life-threatening battle against Voldemort on the other. In the end, the all-powerful and pitiless Voldemort is overcome by the only powers he does not possess: love and friendship.

The *Harry Potter* books articulate in imagination what the intellect hides. This, too, is part of their theme: Voldemort is invisible and hides where he is least expected. In the end, part of him is even found in Harry. This painful realisation becomes the key to Harry's success. The message is that we too must go through the painful realisation of how deeply the intellect has penetrated our thinking and confront the paradigm where it is most powerful: in ourselves.

Pullman's *His Dark Materials* points to similar realities. His heroine Lyra is in possession of an alethiometer, a symbol-reader (and a metaphor for the imagination) that discloses the truth. But unlike those who possessed the compass before her, Lyra needs no books to decipher its pictorial instructions. Her reading is immediate and intuitive. The symbols become alive. Her imagination is the tool that helps her understand her world and navigate its realities.

Her challenger is the Magisterium, an institution seeking total control by severing children from their protecting daemon (their essential self), making them into biddable puppets in a controlled world. Lyra's world is not our own. And yet in it, as in ours, children are attacked by the powerful: the paradigm that seeks total control. Lyra saves herself and others with the help of her imagination.

The popularity of these works shows that on the level of imagination the public psyche knows what public intellectuality hides. It is time to understand what the imagination is communicating: it is the same message that nature articulates by means of environmental crisis and that bees are trying to tell us through their collapse.

Unless we begin to heed these promptings, the intellect will continue to ruin the world. The global psyche will stay in a state of schizophrenia, split between what we unconsciously feel and what we consciously deny, imaginatively know and intellectually discredit.

This split opens the ever-widening gap through which species disappear, the abyss that hinders the crossover of compassion. This is the incision in the mind that allowed Huber to experiment with insemination and Dzierzon to start trading queens. This is the gap that allows beekeepers who love their bees to treat them cruelly, that makes parents expose their children to inappropriate technology and exchange mothers for machines. This is the same gap that allows geneticists to exercise godlike power without godlike care.

The first step is to bridge the gap by taking the imagination seriously: seeing imagination in reality and reality in imagination empowers us to counter brutal rationales with compassion and find the essential in the labyrinth of facts. The second step is to close the gap by means of compassionate ecology. This requires that the imagination penetrates science (the process that Goethe began) and that science enters imagination and what lies beyond (as Steiner attempted). I suggest we begin where Harry Potter ended: with the necessary realisation that the paradigm ruining the world is also in us; that much of the way we think is within that framework; that many matter-of-fact truths are in fact only temporary interpretations and that our attitudes are tempered by centuries of habituation. From there we can make our way to new ideas and use the active imagination to rebuild the hive of civilisation.

22

GLOBAL HIVE

The world is not dangerous because of those who do harm but because of those who look at it without doing anything.

Albert Einstein

One can feel powerless confronted with colony collapse. Economic pressures and pervasive paradigms conspire against the survival of the bee. Political action does not match the speed of decline. My only hope is in the imagination that mobilises the global economist in each of us, the bold, marginalised thinker we all harbour inside ourselves, and the undiscovered politicians that we inevitably are.

In Switzerland recently I was impressed by the high level of direct democracy. There are voting booths everywhere, and everyone is ready to participate in public affairs. The Alpine republic exemplifies civil rule. Minor matters and major affairs are decided upon by a population prepared to shape its own future. Laws are made by the people, for the people. Regulations are localised to suit the situation at hand. I would like to see a voting booth for global concerns, such as the protection of our environment and the salvation of the bee: a means to directly participate in decisions relevant to us all. The environmental crisis is a global concern demanding global action: deforestation of one part of the world affects all others, and rising sea levels encroach on all coasts. Pollution is never just local, and climate change is ubiquitous.

ross borders, our national structures are powerless.

is demands worldwide eco-democracy. A global voting

for planetary concerns would provide the means to engage the economist, politician and visionary in every one of us.

To the intellectual mindset this might seem impossible on logistical and political grounds. There is no infrastructure in place to deal with the administrative complexities, nor is there a unified political body to implement the results. But what seems utopian to the intellect is well within reach for the imagination.

To the imaginal mind, eco-democracy is not only possible, it is already established. The global voting booths are already there. The infrastructure is in place and well equipped to accommodate all levels of civil participation. I am talking about the most widespread, effective, sophisticated and powerful of all voting booths: the cash register. This is the global voting booth, an unequalled tool with which we can take responsibility and vote for the better or worse of this planet.

Here, our daily choices have immediate effect. Every one of our decisions subtly alters the world. We can buy coffee or fair-trade coffee, or coffee that is fair trade and organic or even biodynamic. Every purchase is a vote. Our money is our most political tool. What we spend flows back to the product's origins, and contributes to the proper or improper treatment of land, of workers, farmers, societies. With our daily decisions we endorse better or worse ways of transport, more or less trustworthy companies, wholesalers and retailers. We say yes or no to artificial fertilisers, herbicides and pesticides. We prevent or promote sustainable farming, support or neglect fair returns. We have a hand in what happens to the land, the seas, the air: we put our ten cents' worth of opinion on one side or another of the global scale.

The cashier registers our global care. It is our thermostat of environmental awareness. Compared with this, our political choices leave us powerless because they hand responsibility to parties, politicians and governments. The true political arena is the global voting booth. In the parliament of money, our opinions are registered and put into action. Here, every one of our choices changes the world: we join the battle against economics without care, and commerce without conscience.

As consumers we are intimately connected to the world. This tin made in Poland contains sardines from Norway, olive oil from Italy, garlic from Spain and spices from all over the world. Every week our shopping trolleys fill with olives from Portugal, cheeses from Holland, sugar from Brazil, butter from New Zealand, tea from Taiwan and rice from India. What we purchase in our home towns today becomes reality in Indonesia tomorrow. Every shopping trip is a tour around the world, every meal a culinary circumnavigation of the earth. It is the same with all products. Wool from Australia may be spun in England, dyed in Italy and manufactured in China. Complexity increases when we step from simple products to elaborate machinery. Everything comes from everywhere. Every car is an assembly of the world. 'Made in China' is a partial truth, 'Made on Earth' the complete reality.

This makes conscious consumption a worldwide feedback loop, while thoughtless consumption tightens a noose around the neck of the earth. Whatever we choose, we support. A conventional product may be farmed without regard for the earth: it may deplete the topsoil, spoil the water, pollute the air, diminish biodiversity, impact on forests. It may have travelled halfway around the world, accumulated unnecessary food-miles, wasted fuel and lost much of its freshness. When we buy a burger from a food chain, we salivate on unsavoury practices, social exploitation, monoculture, artificial fertilisers and pesticides that burden our stomach as they burden the earth. Through every financial transaction we become poison or nutrient for the earth, engage ourselves in monoculture or diversity, suppression or liberation. We need to ask the compassionate questions. Is this spoonful of honey poison for the world? Does this jar seal the destiny of bees? Will saving these forty cents pollute a river? Will this additional cost sustain topsoil, this cheque save a forest, this transaction counter climate change?

The answer to these questions must not remain abstract. Knowledge may stir our conscience but not alter our action. The mindset that allows beekeepers who love their bees to treat them cruelly will likely encourage us to continue with consumption without care. What is needed is economic imagination. The more accurately we imagine pesticides penetrating the soil, polluting the groundwater and entering plants, bugs, bees, birds and beasts, the more we feel responsible. The moment I imagine in detail, I am connected. And the moment I feel connected, I care: conscience turns into compassion, and compassion into action.

This picture can be developed further: we can consider the circulation of goods as a metaphor for the circulation of our planet's lifeblood. In this picture we become the perceptive heart mediating between what we take and what we give. Products lose abstraction if we can see, feel and sense them all the way back to their origin. We need to imagine the money for a bottle of milk flowing back to the udder of the cow, to the farmer and the land he cares for, the soil he treats, the landscape he maintains, the culture he upholds. And we need to feel ourselves as part of this money flow and all its effects. The taste on our tongue is the lesser part of our transaction. What matters is how our actions taste to the world. While it is important to buy healthy food for our well-being, it is more important to buy it for the benefit of the earth. When we consider the latter, we become the heart of the global economic circulation, the sense organ that maintains the world.

This awareness is the morality we need to maintain our planet. In the Middle Ages, morality centred around synagogues, mosques, churches. There were few choices and everyone lived, worked, prayed and died inside a close circle of circumstance. The world is local no more. Every one of our actions has worldwide effects. Morality is in the marketplace. The department store is the cathedral and the shopping mall the congregation hall.

Without imagination, consumption is ignorant egoism, a selfish and ultimately destructive cult. With imagination, the necessity of self-care becomes the opportunity to care for the world, and the shopping mall of consumption transforms into a global hive.

23

ECOLIBRIUM

Many hands make light work.

Proverb

Modern economic productivity rests on the division of labour. Our surplus results from the careful distribution of tasks combined with high levels of collaboration. Honey is clearly derived from the same principle working within nature. The industrial evolution of bees demonstrates this step by step.

There are many thousands of varieties of wild bees, most of them solitary, whose existence exhausts them in the small sphere of their own needs. The life of *Prosopis*, for example, is simple and solitary. This basic bee exudes no wax and accumulates only scant stores. Lacking hairs, tufts, brushes and baskets to gather pollen, she swallows the pollen she needs and bores holes into wood or digs into the earth to build a few awkward cells. Other solitary bees such as *Xylocopa* gather temporarily to survive winter. They hang together on a stalk but share no community otherwise. *Dasypoda* form large enduring colonies, but remain solitary in all other habits. As each one fends for itself, their ability to gather honey and service the environment remains slight. With *Panurgi* there emerges a hive with a common entrance and passages.

Bumblebees take the next step. Initially, this gentle giant lives alone.

The fertilised queen builds shapeless wax cells to store honey, pollen and brood. When her daughters hatch, a small state forms around the matriarch. With the increase in community, the construction of cells becomes more elaborate. As work is shared, stores begin to accumulate and ecological impacts increase. But the achievements of the bumblebee remain modest. The colony never exceeds a few hundred inhabitants, collaboration remains rudimentary, and social life often slides into cannibalism and infanticide. Lacking organisational coherence, colony prosperity and ecological effectiveness remain limited.

Both rise dramatically with domesticated bees and peak in *Apis mellifera* – a positive indicator of how culture and nature can mutually enhance each other. Here, one queen resides over a complex civilisation of female workers and drones. Colonies contain up to 60,000 bees working in collaboration. The hexagonal cell, a masterpiece of instinctual design, is fully accomplished and artfully fashioned into comb. Division of labour is complex and varied. Worker bees graduate to various stages, starting as domestic cleaners, then in turn becoming nurse bees, ladies-in-waiting, air-conditioning personnel, watchwomen, foragers, wax makers, comb builders and finally water bearers, the last task in a bee's busy life.

At this level of complexity, the production of honey and the beneficial environmental impact reach their greatest heights. This industrial evolution of bees demonstrates that economy and ecology are two sides of the same coin. Whatever a hive does to further its own prosperity simultaneously furthers that of the environment. The tangible outcome of all this economical–ecological activity is honey. Honey results from the modern economic principle already at work in nature. It is surplus made visible, communal effort become substance, work turned sweet. It is a symbol for surplus achieved in and through community. Honey is prosperity made manifest.

As it is with bees, so it is with humans. A single survivor is hard-pressed. A small, loosely linked community produces little; an organised village more. Fifty people working together will surpass the production of a hundred people toiling on their own. Division of labour allows those suited best for a task to perform it. Doing what they do best serves both the community and themselves: a skilled potter will produce more and better pottery than a smith would. A good smith will shoe a horse faster than a potter could. Everyone profits from everyone else. There are more pots for more people for a cheaper price, better horseshoes, safer horses, more reliable transport and better trade in turn. The effects multiply and prosperity rises as a result.

Modern economic systems have long since reached beehive complexity, yet they lag far behind the perfection of the insect state. The economic crisis shows our shortcoming in society, the ecological crisis our failure in relation to nature. Economic progress seems to come at the expense of ecology.

What makes us fail where bees succeed? As always, it is matter of paradigm. A mindset capable of destroying beehives – the most successful economic organism in nature – is ill-equipped to maintain its own economy. The limited intellect sees limited resources. Scarcity creates competition, and competition creates more scarcity and so on.

In nature, competition certainly plays an important – if circumscribed – part. The single lion will eat the single gazelle, while the lion species (the lion in all lions) will not threaten the existence of the gazelle species. Experiments in which all predators were removed from the Kaibab Plateau in northern Arizona, in order to see the effect on the deer population, demonstrated the interdependence of predator and prey.[1] For a while deer numbers did indeed increase, only to collapse beyond initial numbers later on. The predator that culled the population also kept it at a sustainable, healthy size. In nature, competition and collaboration are finely balanced so that competition serves collaboration.

Nature is competitive in its parts (the single wolf will eat the single deer), but collaborative in totalities (the wolf species supports the deer species). The intellect has elevated the principle operative within parts to a governing law. Applied in the wrong place, where collaboration is the relevant principle, it can only wreak harm on human affairs, the economy, and nature.

Every thriving beehive demonstrates that there is another way; every dying colony reveals how far we are from following it. Contrary to popular belief, the economic principle is in fact one of mutual profit. In economic culture, no one prospers on their own. All profit is inevitably made with the help of others. Our food, clothes, tools, machines, elaborate technology and social services are all based on the same give and take that governs ecology: the continuous, mutually supportive exchange that ensures that the gift of everyone becomes available to everyone else.

There is enough food to feed the world, but not enough imagination to put it on every plate. There is enough profit for everyone to prosper, but not enough compassion to share it around. We do not lack creative solutions, only the mindset to put them in place. The real scarcity is the scarcity of imagination and the paucity of paradigms. Only the imagination can turn problems into possibilities, consumerism into care, competition into collaboration. To the intellect, ecology and economy are competing opposites. To the imagination, they are the two sides of the same ecolibrium: a dynamic, constantly evolving balance between human needs and the needs of nature, a constant fair exchange, a continuous and mutually supportive give and take. The end of one exchange is the beginning of another. The brick that someone produces

Figure 14. *Ecolibrium Stamp,* Horst Kornberger, 2003.

is the start of someone else's house. The money I pay for a loaf of bread is the leaven for the baker's dough. Technology I purchase today will help improve it tomorrow. I have tried to express this mutuality in the *Ecolibrium Stamp* (see Figure 14), a conceptual artwork capturing the underlying perpetuum mobile of love that keeps both nature and culture alive.

Mutually supportive exchange links economy with ecology, culture with nature. In nature, one species supports the other. In culture, each individual is a species in its own right whose decent survival is the fittest task our civilisation can set itself. To turn this idea into reality, we need both personal and communal imagination: personal imagination to extend our self-care through compassion, and communal imagination to find the social forms suitable to our global needs.

24

THE HONEY DOCTRINE

Basically, my sculptures, too, are a kind of Apis cult.

Joseph Beuys

As far as apiculture is concerned there are some hopeful signs of change. Amateur beekeeping is on the rise. Rooftop beekeeping is coming into fashion. New bee-appropriate hives are being designed, and novel and non-violent beekeeping techniques explored. For instance, the Warré hive, developed in France by Abbé Emile Warré in the early part of the twentieth century, is a bee-friendly top-bar hive that allows undisturbed bees to build their own comb. Surplus honey alone is taken. Organic and biodynamic beekeeping explore new avenues of collaboration with hives.

Among the many inspiring exemplars, the bee sanctuary founded by Gunther Hauk in Spikenard Farm in the United States stands out. His biodynamic farm – a mix of refuge, research laboratory and sanctuary – is built around the well-being of bees. Situated in the middle of Monsanto country (that is, country that has been agro-chemically abused by Monsanto, a major producer of chemicals), this initiative serves as an oasis for bees. Hauk experiments with various hive shapes and materials, and I am particularly taken by his rounded bee-appropriate straw and clay hives. Hauk describes his work in *Toward Saving the Honeybee*,[1] a book I wholeheartedly recommend. His work adds a sacred dimension to practical care. Without this dimension our environmental measures will inevitably fall short. The time has come to resurrect the constructively sacred from the destructively profane.

This resurrection is the beekeeping in which we are all involved. Today, responsibility for *Apis mellifera* rests with everyone. Global beekeeping depends on all of us practising compassionate ecology. It is up to us to develop empathy, swarm our paradigms and build a global hive. We are all called to engage our imagination and restore the relatedness we have lost. A potent example of lost relatedness is the famous *Madhu-vidya*, or honey doctrine. This text from the ancient Indian Upanishads testifies to the status of honey as universal binding agent:

> This earth is the honey of all beings, and all beings are the honey
> of this earth. Likewise this bright, immortal person in this earth,
> and that bright immortal person incorporated in the body (are
> both honey.) He indeed is the same as that Self, that Immortal,
> that Brahman, that All.
>
> This water is the honey of all beings, and all beings are the
> honey of this water. Likewise this bright, immortal person in
> this water, and that bright, immortal person, existing as seed in
> the body (both are Madhu). He indeed is the same as that Self,
> that Immortal, that Brahman, that All.
>
> This fire is the honey of all beings, and all beings are the honey
> of this fire...

> After fire follows air, sun, space, moon, lightning, thunder, ether,
> the law, the true, mankind and finally the Self.

> This Self is the honey of all beings, and all beings are the honey of
> this Self...[2]

In this doctrine, ecology and sacred knowledge coincide. The *Madhu-vidya* addresses honey as a unifying agent, a substance shared by otherwise separated parts. It invokes honey as the essence of essences, related to the all-encompassing, all-pervading, omnipresent deity Brahma and its reflection in the individual soul (Atman). It is this underlying unity that relates the earth to all beings and all beings to the earth. Honey was to Indian sages what imagination was to Goethe: the means that binds isolated phenomena into a whole and so closes the gap between subject and object, individual and universal mind. The active imagination is the honey doctrine for our time and Goethe's science a new, contemporary yoga of ecology.

Our intellectual age tends to understand the mystical honey of the *Madhu-vidya* as a subjective metaphor, but to ancient ecology, tangible honey and 'mystical honey' were linked. Through the bee crisis, this link is becoming apparent once more. For today, honey is again, as it was then, the barometer of our relatedness. The potential loss of 'real' honey in our time is a sure indicator of the equally real loss of metaphoric honey: the lack of imagination and relatedness. The result is a paucity of compassion and an inability to connect to the 'bee in all bees'.

To reverse this trend we need a new honey flow between parts and whole, essence and manifestation, science and art, intellect and imagination, individual and society. German artist Joseph Beuys articulated this need in his *Honeypump* at Documenta 6 in Kassel (a major international art event in 1977). His installation comprised two motors pushing honey through a network of pipelines that ran along the entire museum space. *Honeypump* served as backdrop for a series of talks, discussions and workshops in which the artist engaged with the audience to effect social change. Real conversation is inevitably a honey process: it links separated minds by the common agent of active understanding. It touches on the essential dimension of meaning, the true source of transformation. Beuys' work inspired my own engagement with honey. Artworks like *Buddha in Honey* (see Figure 11, p. 84), *The Alphabet of Bees* (see Figure 15, overleaf) and *Honeyclock* (see Figure 16, p. 131) and elaborate on the same theme.

Figure 15. *The Alphabet of Bees: Codex for a new Ecology,* Horst Kornberger, 1987.

Figure 16. *Honeyclock*, Horst Kornberger, 2001.

Following the thread of honey further, I would like to share a new imaginative artwork: *Honey Chapel*, an imaginary monument to compassionate ecology.[3] It consists of a large, simple space containing nothing but a bowl into which a thread of honey falls from the ceiling. Honey falls in a thin thread that spins at its point of contact. It has the amazing ability to remain whole and connected even in extreme dilution. It exemplifies in substance the connectivity that marks the hive and is the distinguishing feature of the imagination: it is the obvious substance to link what otherwise remains separated. The falling honey is not sealed away, but experienced as visible, tangible, fragrant substance. I imagine the space dimly lit and the honey artfully illuminated. From the bowl, it is noiselessly returned via a solar powered pump to fall in perpetual motion. I see this thin line of honey as a contemporary Ariadne-thread, a vertical link that connects the labyrinth of partial knowledge with the unifying dimension of compassionate, essential knowing: a golden liquid stitch linking isolated concepts into comprehensive pictures, reconnecting separated science through unified imaginations.

In a moving essay, molecular scientist and beekeeper Johannes Wirz sees the flight paths of bees as threads weaving a gigantic tapestry around the globe.[4] I see the active imagination doing the same. Pollinating the global mind, it weaves a new 'eco-sphere of meaning' around the future of this globe, combining the vertical warp of the imagination with the horizontal weft of bees. Together they make the fabric of the future.

The bee crisis is a call to make the care of the earth the unifying feature of our time. *Honey Chapel* is an attempt to articulate this unity by expressing an ecomorality beyond philosophical and religious outlooks, for the language of honey is suited to every tongue, and the symbolism of honey understood in every culture.

I therefore see *Honey Chapel* as a temple of global environmentalism and a platform for the paradigm change we so desperately need. I imagine honey chapels at all the cultural acupuncture points of this planet: in the Pantheon in Rome, the Acropolis in Athens, in Egyptian temples. I can see a golden stream of honey artfully lit in the dim interior of Chartres Cathedral adding fluidity to imaginations cast in stone. I can see honey falling through the open space of the Guggenheim Museum in New York, adding the vertical to spiralling passages of art. I can see honey thread linking contributions of Steiner, Goethe and Beuys in the Goetheanum, Switzerland. I can see honey falling through the open air in Borobudur, covering the statues of Buddha with the fluid gleam of environmental care. I see it in the UN headquarters as a reminder that unity of the world is impossible without unifying imaginations. I can see it in old temples and new places of worship as a pointer that the care of the earth is the unifying religion of our times, and in museums as a reminder that ecology is everyone's contribution to the art of maintaining the planet. I see it wherever human beings meet to converse, to collaborate with one another and with nature; I see it wherever concerns are worldwide and imaginations real.

AFTERWORD

I have tried to show how Steiner's intuition that ongoing manipulation of queen bees would lead to a worldwide collapse of bee colonies can be readily understood with common sense and a minimum of empathetic sensitivity – if we are willing to employ it against the promptings of profit and scientific indifference. While many factors affect the well-being of bees, it is the torturous treatment of the queen bees that needs to be stopped first and foremost.

With this in mind, I will make a suggestion that may seem radical at first but is only reasonable in the light of a crisis that endangers global food production and hence human civilisation.

I propose an internationally binding law that recognises each beehive and its resident queen as a legal entity with personhood status. As such, it is entitled to live and express itself according to its own nature, defined not by economic imperatives but through empathetic observation of what is essential to bees. I am calling, in other words, for a worldwide ban on all torturous anti-bee procedures as a necessary step towards a global eco-democracy. The only way to win the war against our own extinction is to make peace with nature. The movement to grant mountains, forests and rivers legal rights is already on the way. Ecuador has codified the rights of nature; in New Zealand, Mount Taranaki and the Whanganui river have been granted legal personhood. India has done the same to the rivers Ganges and Yamuna, and Colombia to the Atrato.

Bees, as representatives of nature within nature, as master ecologists and economists, as creators of societies whose dynamic complexity we are far from being able to emulate, ought to be the first animate members of a planetary protection act that recognises the rights of each species to self-express.

When I first wrote this book six years ago, I put my hope into the paradigm changes, the global voting booths, alternative beekeepers and bee-aware groups all around the world. Today, I know this is not enough. Immediate political action at all levels is needed. The bee crisis is a global emergency that needs global action beyond petty political concerns and party agendas. My hope is that politicians will rise to their responsibility, that crisis-aware nations will inspire others, and that the UN will provide an eco-democratic vision that is meaningful to everyone because it is important to all. Without such measures, we endanger not only bees, but ourselves.

By taking the first courageous steps toward the preservation of the bee, we may yet hope to turn this ailing planet into a global hive.

ENDNOTES

1. From Beuys to Bees

1. *Honeyclock* (2001) was my contribution to the TIME-EMIT (2001) exhibition at the Moore's Building Contemporary Art Gallery in Fremantle, with Patrick and Loman McCann and long-time friend and art-collaborator Tom Mùller. *Honeyclock* is a timepiece that measures quality rather than quantity of time. The clock was completed with the help of master woodworker Nisargam Wichtermann and glassblower Dennis Clair. The workings of the clock can be viewed at www.horstkornberger.com.

2. *The Bee Master* is a four-day festival drama written by Jennifer Kornberger in collaboration with West Australian composer Paul Lawrence. The play contained in metaphoric form many of the themes elaborated in these pages.

3. Rudolf Steiner, *Bees*, 8 lectures (GA 351), Dornach, Trans. T. Braatz, Great Barrrington, NY, 1998, p.178. Two questions answered by Dr Rudolf Steiner after lectures to workmen on 10 December 1923.

4. Ibid.

8. Microscope and Mind

1. The illustration first appeared in a collection of satires translated by Stelluti for Cardinal Barberini, whose family coat of arms contained three bees (see David Bardell, 'The First Records of Microscopic Observations', *BioScience*, Vol. 33, No. 1, January 1983, pp.36–38).

2. Maurice Maeterlinck intuits this relationship in his 1901 classic *The Life of the Bee*, London, George Allen & Unwin, 1930, p.3.

11. Macroscope

1. Jan Dzierzon, *Dzierzon's Rational Beekeeping: Or the Theory and Practice of Dr Dzierzon*, Whitefish, MT, Kessinger Publishing, 2010.
2. In 1660, Robert Boyle (1627–91) published *New Experiments Physico-Mechanicall: Touching the Spring of the Air, and its Effects...*. In it, he described his experiment showing the effect of reduced air pressure on a bird.
3. Proponents include Jochen Bockemühl, Henri Bortoft, John Davy, Stephen Edelglass, Hans Gebert, Craig Holdrege, Wolfgang Schad, Theodor Schwenk, Georg Maier and many others.

12. Goethe: The Apprentice of Nature

1. The *os intermaxillare* was a bone in mammals that was believed not to exist in human beings, and its supposed absence was taken as proof that human beings are fundamentally different from mammals.
2. Rudolf Steiner, *Goethean Science*, Spring Valley, NY, Mercury Press, 1988, S.108.
3. Goethe, *Faust*, New York, Norton Critical Edition, 1976, Part I, Scene 6, Act 2, lines 69–74.
4. Some believe this statement is unrepresentative as Bacon also said much to the contrary. I agree that this quote is not typical of the Renaissance philosopher. I use it here not because it exemplifies his wide-ranging musings, but because it characterises his science. Bacon's experimental and statistical methods cannot but put nature on the rack and wrest her secrets away. I believe the quote gained unmerited popularity because it exposes the fundamentally antipathetic element in modern Baconian science.
5. Henri Bortoft, *The Wholeness of Nature: Goethe's Way of Science*, Edinburgh, Floris Books, 1996.

13. Compassionate Ecology

1. Cited in K.R. Biermann (ed.), *Alexander von Humboldt. Aus meinem Leben. Autobiographische Bekenntnisse*, Leipzig, Urania Verlag, 1989, p. 180.

2. To know time intimately requires the capacity to know it on its own terms. Steiner suggests that we can approach time in the way that Goethe approached plants: we can become familiar with the way the future unfolds into the present by imagining an exact sequence of events in reverse order. This loosens the chain of cause and effect that ties us to habitual perceptions, without altering the sequence. The aim is to perceive in pictorial wholeness what the intellect separates into parts. The result is the ability to perceive time in terms of wholeness and to better anticipate how our present actions will affect the future.

14. From Conscience to Compassion

1. Orestes is the son of Agamemnon and Clytemnestra. When Agamemnon returns after the sack of Troy, Clytemnestra and her lover Aegisthus kill him immediately. This puts Orestes in a tragic situation: by divine law he is obliged to avenge his father, but in doing so must kill his mother (which is a sin).

   When he kills his mother he is relentlessly pursued by the Erinyes, vengeful spirits that drive him to the edge of madness. In Aeschylus's earlier version of the play, the Erinyes appear as powerful goddesses on stage. In Euripides's later version, the Erinyes have metamorphosed into psychological realities. The outer has become the inner. The goddesses have become conscience.

15. Global Empathy

1. Heinrich Harrer, *Seven Years in Tibet*, New York, NY, Jeremy P. Tarcher/Putnam, 1997, p.188.

16. Beehive Metaphors

1. I am indebted to my friend John Stubley for pointing me toward the work of the late David Mowaljarlai, a Ngarinyin lawman and elder, who movingly describes land-care in the north of Western Australia (see David Mowaljarlai and Jutta Malnic, *Yorro Yorro*, Rochester, VT, Inner Traditions, 1993).

2. Quoted from the Salt Magical Papyrus, in Hilda M. Ransome, *The Sacred Bee in Ancient Times and Folklore*, Mineola, NY, Dover Publications, 2004, p.33.

3. Aristotle, *On the Generation of Animals*, Kypros Press, Chapter 10.

17. Bee Frames and Mind Frames

1. Owen Barfield, *Saving the Appearances: A Study in Idolatry*, Hanover, NH, Wesleyan University Press, 1988.

2. Horst Kornberger, *The Power of Stories: Nurturing Children's Imagination and Consciousness*, Edinburgh, Floris Books, 2008, Chapters 1 to 7.

3. In Egyptian myth, for instance, death is a pervasive theme. Creator god Ra dies. His heir Osiris dies twice and his dismembered body is buried all over Egypt. Egyptian civilisation developed elaborate rituals around death and burial, and eventually followed its myth into the grave. Almost all we know of it is excavated from burial sites. Indian culture, by contrast, is marked by the frequent intervention of the creator god Vishnu. In the form of his avatars he continuously saves the world from destruction. Indian culture accordingly has again and again risen like a phoenix from the vicissitudes of time, conquest and oppression.

   Greek myths such as Theseus and the Minotaur, Perseus and Medusa and the Battle with the Centaurs prepared the scientific and philosophic achievement centuries before they came about. These tales portray the pre-intellectual 'wild mind' in monstrous mixtures of man and beast, and the emerging intellect through the heroes who overcome them. Theseus escaped the labyrinth of the Minotaur with the help of a thread (the thread of thought), Perseus avoids the petrifying stare of the Medusa with the help of his polished silver shield (by means of reflection) and Pirithous gains victory over intoxicated centaurs carried away by passion with the help of the god Apollo (the serene mind unaffected by instinct).

18. The Choreography of Care

1. Rudolf Steiner, *Bees*, 8 lectures (GA 351), Dornach, Trans. T. Braatz, Great Barrington, NY, 1998. Lecture II.

2. William H. James, 'The Variations of Human Sex Ratio at Birth During and After Wars, and Their Potential Explanations', *Journal of Theoretical Biology*, Vol. 257, No. 1, March 2009, pp.116–232.

19. Swarming Paradigms

1. Brian Keats, 'Australian Native Bees and the European Honey Bee', July 2008, https://www.astro-calendar.com/shtml/Research/indigenousbees.shtml.
2. If we take ecology seriously and assume wholeness beyond separation of mind and nature, we can expect that these capacities must be there, for the global environment regulates the production of paradigm-changing capabilities with the same necessity with which it regulates the productivity of the hive through swarming.

20. Compassion Collapse Disorder

1. Georg Kühlewind, *The Logos-structure of the World: Language as a Model of Reality*, Great Barrington, MA, Lindisfarne Books, 1992 and *Star Children: Understanding Children Who Set Us Special Tasks and Challenges*, Forest Row, Temple Lodge, 2004.
2. Ken Robinson, 'Do Schools Kill Creativity?', TED Talk, 2006, https://www.ted.com/speakers/sir_ken_robinson. Sir Ken Robinson conducted the study about the creative use of paperclips for the Royal Society of Medicine.
3. The problem of genetic manipulation has been brilliantly explored by American biologist Craig Holdrege in *Genetics and the Manipulation of Life: The Forgotten Factor of Context*, Hudson, NY, Lindisfarne Press, 1996.

23. Ecolibrium

1. Clay McCulloch, 'A History of Predator Control and Deer Productivity in Northern Arizona', *The Southwestern Naturalist*, Vol. 31, No. 2, pp. 215–220.

24. The Honey Doctrine

1. Gunther Hauk, *Toward Saving the Honeybee*, San Francisco, Biodynamic Farming and Gardening Association, 2002.
2. *The Upanishads*, Part 2, http://www.sacred-texts.com/hin/sbe15/sbe15098.htm.
3. The idea of a honey thread falling through space was initially conceived after the TIME-EMIT exhibition in 2001, in collaboration with artist friend Tom Mùller.
4. Johannes Wirz, 'Golden Threads and the Golden Fleece', in Taggart Siegel and Jon Betz (eds), *Queen of the Sun: What are the Bees Telling Us?*, Forest Row, Clairview.

BIBLIOGRAPHY

Adams, David, 'The Artistic Alchemy of Joseph Beuys' in Steiner, Rudolf, *Bees*, 8 lectures (GA 351), Dornach, Trans. T. Braatz, Great Barrington, NY, 1998

Apiculture Program, North Carolina State University, 'The Development of the Honey Bee Instrumental Insemination', *Beekeeping Note* 2.14, https://wncbees.org/wp-content/uploads/2014/07/The-Development-of-Honey-Bee-Instrumental-Insemination-NCSU-2.14.pdf

Aristotle, *On the Generation of Animals*, Kypros Press

Bardell, David, 'The First Records of Microscopic Observations', *BioScience*, Vol. 33, No. 1, January 1983, pp.36–38

Barfield, Owen, *Saving the Appearances: A Study in Idolatry*, Hanover, NH, Wesleyan University Press, 1988

—, *Speaker's Meaning*, Middletown, CT, Wesleyan University Press, 1984

—, *History in English Words*, Hudson, NY, Lindisfarne Press, 1967

—, *Romanticism Comes of Age*, Middletown, CT, Wesleyan University Press, 1986

Barton, Matthew, 'What Can We Do for Bees', *New View*, 2010

Bethmann, Dirk and Kvasnicka, Michael, 'Why Are More Boys Born During War? Evidence from Germany at the Midcentury', Ruhr Economic Paper No. 154, https://ssrn.com/abstract=1514383

Biermann, K.R. (ed.), *Alexander von Humboldt. Aus meinem Leben. Autobiographische Bekenntnisse*, Leipzig, Urania Verlag, 1989

Bockemühl, Jochen, *In Partnership with Nature*, Wyoming, RI, Bio-Dynamic Literature, 1981

— (ed.), *Toward a Phenomenology of the Etheric World*, Spring Valley, NY, Anthroposophic Press, 1985

Bortoft, Henri, *The Wholeness of Nature: Goethe's Way of Science*, Edinburgh, Floris Books, 1996

Boyle, Robert, *New Experiments Physico-Mechanicall, Touching the Spring of the Air, and Its Effects*, 1660

Bulfinch, Thomas, *Bulfinch's Mythology*, New York, NY, Modern Library, 1998

Chapman-Taylor, R. and Davy, I., *Practical Beekeeping: Handbook for Australia and New Zealand*, Melbourne, Inkata Press, 1988

Colquoun, Margaret and Ewald, Axel, *New Eyes for Plants*, Stroud, Hawthorn Press, 1996

Davy, John, *On Hope, Evolution and Change: Selected Essays*, Stroud, Hawthorn Press, 1985

Dzierzon, Jan, *Dzierzon's Rational Beekeeping: Or the Theory and Practice of Dr Dzierzon*, Whitefish, MT, Kessinger Publishing, 2010

Ede, Piers Moore, *Honey and Dust: Travels in Search of Sweetness*, London, Bloomsbury, 2005

Edelglass, Stephen et al., *Matter and Mind: Imaginative Participation in Science*, Edinburgh, Floris Books, 1992

Emerson, Ralph Waldo, *Nature: Selected Writings*, New York, NY, Signet Classic

Ende, Michael, *The Neverending Story*, London, Penguin, 1984

Glenn, Tom, 'Commercial Use of Instrumental Insemination', *Apimondia*, Melbourne, 2007, http://www.glenn-apiaries.com/apimondia_1.html

Goethe, Johann Wolfgang, *Faust*, New York, NY, Norton Critical Edition, 1976

—, *The Metamorphosis of Plants*, Cambridge, MA, Massachusetts Institute of Technology, 2009

Graves, Robert, *The Greek Myths*, London, Penguin, 1992

Grimal, Pierre (ed.), *Larousse World Mythology*, London, Hamlyn Publishing Group, 1973

Guth, Steven, *Honeybee*, Milton, Queensland, Jacaranda Press, 1976

Harbo, John R. and Rinderer, Thomas E., 'Breeding and Genetics of Honey Bees', in *Beekeeping in the United States*, Agriculture Handbook No. 335, Washington, DC, 1980

Harrer, Heinrich, *Seven Years in Tibet*, New York, NY, Jeremy P. Tarcher/Putnam, 1997

Harris, J.W. and Harbo, J.R., 'Changes in Reproduction of Varroa Mites After Honey Bee Queens Were Exchanged Between Resistant and Susceptible Colonies, *Apidologie*, Vol. 31, Edition 6, 2000, pp.689–699

Hauk, Gunther, *Toward Saving the Honeybee*, San Francisco, CA, Biodynamic Farming and Gardening Association, 2002

Heaf, David, 'Sustainable Bee-Friendly Beekeeping', *New View*, http://www.bee-friendly.co.uk

Holdrege, Craig, *Genetics and the Manipulation of Life: The Forgotten Factor of Context*, Hudson, NY, Lindisfarne Press, 1996

Huber, François, *New Observations on the Natural History of Bees*, London, 1806

James, William H., 'The Variations of Human Sex Ratio at Birth During and After Wars, and Their Potential Explanations', *Journal of Theoretical Biology*, Vol. 257, No. 1, March 2009, pp.116–232

Johnson, Renée, 'Honey Bee Colony Collapse Disorder', Washington, DC, Congressional Research Service, 2010, https://fas.org/sgp/crs/misc/RL33938.pdf

Johnstone, Mark, 'Rearing Queen Bees', *Primefacts*, No. 828, September 2008, https://www.dpi.nsw.gov.au/__data/assets/pdf_file/0003/252129/Rearing-queen-bees.pdf

Kastner, Jeremy and Wallis, Brian, *Land and Environmental Art*, London, Phaidon, 1988

Keats, Brian, 'Australian Native Bees and the European Honey Bee', July 2008, https://www.astro-calendar.com/shtml/Research/indigenousbees.shtml

Kornberger, Horst, 'Bee Crisis – World Crisis', in Taggart Siegel and Jon Betz (eds), *Queen of the Sun: What are the Bees Telling Us?*, Forest Row, Clairview

—, *The Power of Stories: Nurturing Children's Imagination and Consciousness*, Edinburgh, Floris Books, 2008

—, *The Writer's Passage: A Journey to the Sources of Creativity*, Hilton, Integral Arts Press, 2008

Kornberger, Jennifer, *I Could Be Rain*, Cottesloe, Sunline Press, 2007

Kraetz, Otto, *Alexander von Humbold: Wissenschaftler, Weltbürger, Revolutionär*, München, Callwey, 2000

Krishnananda, Swami, *The Madhu-Vidya* in *The Brihadaranyaka Upanishad* Rishikesh, The Divine Life Society, https://www.swami-krishnananda.org/brhad_00.html

Kühlewind, Georg, *Bewusstseinsstufen*, Zuerich, Verlag Freies Geistesleben, 1980

—, *Die Wahrheit Tun*, Zuerich, Verlag Freies Geistesleben, 1982

—, *The Logos-structure of the World: Language as a Model of Reality*, Great Barrington, MA, Lindisfarne Books, 1992

—, *Star Children: Understanding Children Who Set Us Special Tasks and Challenges*, Forest Row, Temple Lodge, 2004

Leach, Roland, McCauley, Shane and Ward, Donna (eds), *Weighing of the Heart: An Anthology of Emerging West Australian Poets*, Cottesloe, Sunline Press, 2007

Lewis, C. S., *The Chronicles of Narnia*, 1950–56

Lowenstein, Tom, *The Vision of the Buddha: Buddhism – The Path to Spiritual Enlightenment*, London, Duncan Baird Publishers, 2000

Maeterlinck, Maurice, *The Life of the Bee*, London, George Allen & Unwin, 1930

McCulloch, Clay, 'A History of Predator Control and Deer Productivity in Northern Arizona', *The Southwestern Naturalist*, Vol. 31, No. 2, pp.215–220

Mowaljarlai, David and Malnic, Jutta, *Yorro Yorro: Aboriginal Creation and the Renewal of Nature – Rock Paintings and Stories from the Australian Kimberley*, Rochester, VT, Inner Traditions, 1993

Nesfield-Cookson, Bernard, *Rudolf Steiner's Vision of Love: Spiritual Science and the Logic of the Heart*, Wellingborough, Aquarian Press, 1983

Pullman, Phillip, *His Dark Materials*, London, Scholastic, 1995–2001

Pundyk, Grace, *The Honey Spinner*, Millers Point, Murdoch Books, 2008

Ransome, Hilda M., *The Sacred Bee in Ancient Times and Folklore*, Mineola, NY, Dover Publications, 2004

Robinson, Ken, 'Do Schools Kill Creativity?', TED Talk, 2006, https://www.ted.com/speakers/sir_ken_robinson

Root, A.I., *The ABC and XYZ of Bee Culture*, Medina, OH, A.I. Root Company, 1980

Rowling, J.K. *Harry Potter*, London, Bloomsbury, 1997–2007

Schwenk, Theodor, *Sensitive Chaos*, Forest Row, Rudolf Steiner Press, 1990

Sheldrake, Rupert, *Dogs that Know when Their Owners Are Coming Home*, London, Arrow Books, 1999

Siegel, Taggart and Betz, Jon (eds), *Queen of the Sun: What are the Bees Telling Us?*, Forest Row, Clairview, 2011

Snodgrass, Robert E., *The Anatomy of the Honey Bee*, Washington, DC, Government Printing, 1910

Sommerville, Doug, 'Varroa Mites', *Primefacts*, No. 861, January 2009, https://www.dpi.nsw.gov.au/__data/assets/pdf_file/0006/268026/Varroa-mites.pdf

Spence, Michael, *After Capitalism*, Hillsdale, NY, Adonis Press, 2014

Stachelhaus, Heiner, *Joseph Beuys*, München, Wilhelm Heyne Verlag, 1993

Steiner, Rudolf, *Bees*, 8 lectures (GA 351), Dornach, Trans. T. Braatz, Great Barrington, NY, 1998

—, *Education As a Force for Social Change*, Hudson, NY, Anthroposophic Press, 1997

—, *Goethe's World View*, Spring Valley, NY, Mercury Press, 1985

—, *Goethean Science*, Spring Valley, NY, Mercury Press, 1988

—, *Intuitive Thinking As a Spiritual Path*, Hudson, NY, Anthroposophic Press, 1995

—, *Knowledge of Higher Worlds – How Is It Achieved?*, Bristol, Rudolf Steiner Press, 1993

—, *The Boundaries of Natural Science*, Spring Valley, NY, Anthroposophic Press, 1985

Tacey, David J., *Edge of the Sacred: Transformation in Australia*, Sydney, HarperCollins, 1995

Tennyson, G.B. (ed.), *A Barfield Reader*, Edinburgh, Floris Books, 1998

Twist, Lynne with Barker, Teresa, *The Soul of Money: Transforming Your Relationship with Money and Life*, New York, NY, W.W. Norton & Company, 2003

Warnke, Ulrich, *Bees Birds and Mankind: Destroying Nature by 'Electrosmog'*, Kempten, Competence Initiative for the Protection of Humanity, Environment and Democracy, 2009

Weiler, Michael, *Bees and Honey: From Flower to Jar*, Edinburgh, Floris Books, 2006

Wirz, Johannes, 'Golden Threads and the Golden Fleece', in Taggart Siegel and Jon Betz (eds), *Queen of the Sun: What are the Bees Telling Us?*, Forest Row, Clairview

Online Resources

The Nature Institute is a not-for-profit organisation that serves as a forum for the exchange of ideas about the re-visioning of science and technology in an effort to realign humanity with nature: www.natureinstitute.org

The science section at the Goetheanum, Switzerland, is an international research institution that pioneers complementary approaches to the natural sciences inspired by the works of Goethe and Steiner: http://science.goetheanum.org/358.html

Rupert Artificial Insemination of Honey Bees: http://www.vegetus.org/honey/art.htm

Artpendix

All figures are taken from Wikimedia Commons unless otherwise stated and the original source given where possible.

Figure 1. *Bicorp Man*. Rock art, *c.* 6000 BCE, Araña Caves, Valencia, Spain.

Figure 2. *Beehives in the Tomb of Pabasa.* Tomb painting, *c.* 2400 BCE.

Figure 3. Types of bee. From the *Illustrated Dictionary of Gardening: A Practical and Scientific Encyclopedia of Horticulture,* edited by George Nicholson, *c.* 1885.

Figure 4. *Bees,* Francesco Stelluti. Micrograph, 1630. Initially prepared for *Apiarium,* a work on bees by Lincean Academy founder Federico Cesi.

Figure 5. *Ovaries of the Bee Queen,* Jan Swammerdam. Micrograph, 1669. First published in the *Historia Insectorum Generalis (The Natural History of Insects).* Wellcome Collection.

Figure 6. Swiss naturalist *François Huber* (1750–1831).

Figure 7. *Huber Hive (Leaf Hive),* François Huber, 1806. First published in *New Observations on the Natural History of Bees.*

Figure 8. Illustration from *The Anatomy of the Honey Bee,* R.E. Snodgrass, 1910.

Figure 9. *An Experiment on a Bird in the Air Pump,* Joseph Wright of Derby. Oil on canvas, 1768. National Gallery, London.

Figure 10. *Johann Wolfgang von Goethe* (1749–1832).

Figure 11. *Buddha in Honey 2,* Horst Kornberger, 2001. From Icons of the Environmental Age. © Horst Kornberger.

Figure 12. *Gold plaques embossed with winged bee goddesses from Camiros, Rhodes,* seventh century BCE. British Museum.

Figure 13. *Diana (Artemis) of Ephesus statue* in the Candelabra Gallery, Vatican Museums.

Figure 14. *Ecolibrium Stamp,* Horst Kornberger, 2003. From Icons of the Environmental Age. © Horst Kornberger.

Figure 15. *The Alphabet of Bees: Codex for a new Ecology,* Horst Kornberger, 1987. © Horst Kornberger.

Figure 16. *Honeyclock,* Horst Kornberger, 2001. From Icons of the Environmental Age. © Horst Kornberger.

ABOUT THE AUTHOR

Horst Kornberger is an interdisciplinary artist, educator, writer, poet, performer and researcher into the field of imagination and creativity. Horst was born in Austria in 1959 and began his career as a visual artist. He studied Speech and Drama in England, Goethean Studies in the US and Waldorf Education in Australia. Horst is a director of The Writing Connection and Creativity Consultants Worldwide.

He is the author of *Global Hive, The Writer's Passage, The Power of Stories, The Delphi Project* and *Taliesin: Recovering the Poetic Self.*

Horst lectures internationally on themes of creativity, education, ecology and the use of the imagination as a healing and community building tool. Horst's current artistic practice focuses on creating fields of interdisciplinary enquiry. Find out more at:

www.horstkornberger.com
www.thewritingconnection.com.au
www.thetheatreofthesea.net
http://horstkornberger.blogspot.com.au

Lost Knowledge
of the Imagination

Gary Lachman

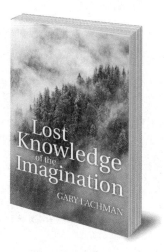

'Very important... Immensely readable.' – Phillip Pullman

'Lachman creates a history of ideas that fascinates and excites.'
– New York Journal of Books

Imagination is a core aspect of being human. Our imagination allows us to fully experience ourselves in relation to the world and reality, and plays a key role in creativity and innovation. Since the seventeenth century, however, imagination has been sidelined and dismissed as 'make believe'.

This insightful and inspiring book argues that we must redress this balance. Ranging from the teachings of ancient mystics to the latest developments in neuroscience, *Lost Knowledge of the Imagination* introduces the reader to a philosophy and tradition that restores imagination to its rightful place, and argues that it is not only essential to our knowing reality to the full, but to our very humanity itself.

 Also available as an eBook

florisbooks.co.uk

Thinking Outside the Brain Box
Why Humans are not Biological Computers

Arie Bos

Does our brain produce consciousness? This is a widely held belief, even amongst scientists, founded on a conception of the world that is purely materialistic. In this worldview the brain is seen as a kind of biological computer.

However, recent developments in neuroscience suggest that our experiences shape the circuits of our brain, and even stimulate the brain to grow. So to an extent, we shape our own brain just through being alive. And it is by means of our brain that we develop as a person and form our 'self', with all its associated significance and values.

In this revealing study of brain, body and consciousness, Arie Bos examines the limitations of the materialist view to explain our human experience. Exploring the ideas of free will and responsibility, he rejects the view that only physical matter determines our thoughts and actions and in doing so opens a door to a wider spiritual reality.

 Also available as an eBook

florisbooks.co.uk

Taking Appearance Seriously
The Dynamic Way of Seeing
in Goethe and European Thought

Henri Bortoft

'A seminal text that demands the widest audience.' – Paradigm Explorer

In this fascinating book, renowned thinker and physicist Henri Bortoft explores a way of seeing that draws attention back from what is experienced, into the act of experiencing. Understood in this way, perception becomes a dynamic event, a 'phenomenon', in which the observer participates.

Bortoft guides us through this dynamic way of seeing in various areas of experience, and discusses how twentieth-century European thinkers radically overturned the notion of dualism.

Expanding the scope of his previous book, *The Wholeness of Nature*, the author shows how Goethean insights combine with the dynamic way of seeing in continental philosophy to offer us an actively experienced 'life of meaning'.

 Also available as an eBook

florisbooks.co.uk

The Secrets of Bees
An Insider's Guide to the Life of Honeybees

Michael Weiler

'A refreshing insight into the bees' world.' – Bee Keeping

'A wonderful journey through the beekeeping year.' – Star and Furrow

Bees make honey; we all know that. But what happens between the bee buzzing around our backyard, and the sticky knife in the jar, is a mystery to most of us. How many bee-hours does it take to make just one jar of honey? What do the honeybees' waggling dances really mean? Why do bees swarm? What is a 'house bee'?

From exploring their life cycle and development, to revealing their societies and behaviour, expert biodynamic beekeeper Michael Weiler answers these questions and many more.

Combining poetic observations with scientific detail, *The Secrets of Bees* uncovers the incredible world of these remarkable insects.

 Also available as an eBook

florisbooks.co.uk